Table of Contents

About the Stories

The 27 stories in *Read and Understand, Science, Grades 2–3* address science objectives drawn from the National Science Education Standards for grades K through 4. There are nonfiction and realistic fiction stories in the areas of life science, physical science, earth & space science, and science & technology.

When dealing with science content, certain specific vocabulary is necessary. This science vocabulary was discounted in determining readability levels for the stories in this book (which progress from low-second to high-third grade). A list of suggested science vocabulary, as well as other challenging words, is provided on pages 3 and 4.

How to Use the Stories

We suggest that you use the stories in this book for shared and guided reading experiences. The stories provide excellent opportunities to teach nonfiction reading skills, such as scanning for information and gleaning information from illustrations and captions.

Prior to reading each story, be sure to introduce the suggested vocabulary on pages 3 and 4.

The Skills Pages

Each story is followed by three pages of activities covering specific skills:

- comprehension
- vocabulary
- a related science or language arts activity

Comprehension activities consist of two types:

- multiple choice
- write the answer

Depending on the ability levels of your students, the activity pages may be done as a group or as independent practice. It is always advantageous to share and discuss answers as a group so that students can learn from peer models.

Vocabulary to Teach

The content of the stories in *Read and Understand, Science, Grades 2–3* requires that specific vocabulary be used. Introduce these words before presenting the story. It is also advisable to read the story to pinpoint additional words your students might not know.

Plants from Seeds
seed, plant, seedcase, splits, shoot, cycle, sunlight, roots, stem, leaves, pods, flowers

A Thunderstorm
clouds, thunderstorm, lightning, flashed, rumble, electricity, expand, raindrops

Scientists Use Tools
scientists, guinea pig, scale, weighed, measured, sore, strange, lab, microscope, science, research

The Moon
orbits, plains, craters, lifeless, astronauts, special, Earth, planet, space, rotates, reflects, phases, oxygen, crescent, gibbous

How Seeds Travel
library, stickers, weeds, fling, feces, vacant lot, puffball, maple, seedpods

Mr. de Mestral's Invention
George de Mestral, invention, inventor, idea, burrs, microscope, fastener, velour, crochet, meters, Velcro®, millions, curiosity

Trees
habitat, climate, swamps, deserts, broadleaf, conifer, palm, chlorophyll, reproduce, trunk, branches, bark, roots, leaves, seeds, absorb, oxygen

Twinkle, Twinkle, Little Star
wonder, diamond, atmosphere, gases, planets, Earth, layer, thousands, million

Magnets
attract, repel, object, magnet, force, magnetic field, poles, opposite, metal, iron, surrounds, temporary magnet

Moth Life Cycle
cycle, metamorphosis, larva, molting, pupa, cocoon, adult, female, caterpillar

Listen for the Sounds
vibrates, vibrations, outer ear, inner ear, unpleasant, pleasant, solids, liquid, gas, music, laughter, echoes

Phil's Science Log
materials, wondered, experiment, conduct, demonstration, logbook

Rainbows
reflected, violet, least, prism, ceiling, arc, raindrops, rainbow, indigo, sunlight

The Earth Is Always Moving
rotates, rotation, orbits, tilt, seasons

Tanisha's Science Project
invited, responsible, friction, hollows, rough, display, objects

Building a Tree House
gathered, complained, simple machines, wheelbarrow, ramp, anxious, pulley

Honeybees
cell, hive, drones, forager, nectar, pollen, waggle dance, round dance, queen bee, guard, intruders, stomach, worker bee

Animals Without a Backbone
backbone, external, fangs, abdomen, spinnerets, antennae, skeleton, common, insects, spider, mosquito

Take a Closer Look
observe, universe, telescope, observatories, microscope, scientists, Saturn, Jupiter, details, human, cells, objects

Keeping Warm
flows, conduction, conductor, insulators, insulation, object

It's Not Just Dirt!
dirt, organic material, decayed matter, organisms, bacteria, silt, clay, sandy soil, resources, tunneling creatures, topsoil, subsoil, combination soil, solid

Platypus
Australia, mammal, warm-blooded, monotremes, kilograms, burrows, snout, spurs, habitat, awkward, crayfish, belly, nostrils, prey, ridges, energy, female, tunnel, leathery, adapted, echidna, centimeters

What Happened to My Pizza?
digestion, energy, digestive system, saliva, esophagus, stomach, small intestine, feces, molecules, large intestine, bloodstream

A North American Desert
Death Valley, Siberia, evaporates, dunes, drought, adapt, predators, dew

Why Recycle?
recycle, recycling, landfills, incinerators, energy, pollution, trash, garbage, materials, useful, reduces, resources, creates, containers, separated

Plastic—A Manmade Product
manmade, molds, molded, liquid, solid foam, pellets, insulation, lightweight, inexpensive, combination

Fossils
Tyrannosaurus Rex, fossil, estimate, ancient, decayed, minerals, seeped, paleontologists, prehistoric, Sue Hendrickson, South Dakota, complete, skeleton, permanent, Chicago, museum, remains, dinosaur, discovery, decomposes

Plants from Seeds

A **seed** is the start of a new plant. Where does the seed come from? How does a seed turn into a new **plant**?

Bean seeds are planted in the ground. A tiny new plant is inside each seed. Water will help the seeds begin to grow. Water and **sunlight** will help the bean plants grow bigger.

First the **seedcase** splits open. Tiny roots start to grow down into the ground. A **shoot** begins to grow up out of the soil.

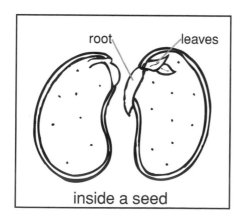

root   leaves

inside a seed

The plant's **stem** grows longer, and many leaves grow on the stem. **Leaves** make food for the plant.

Read and Understand, Science • Grades 2–3 • EMC 3303

Day by day the bean plant grows taller. It has many stems now. The stems are covered with leaves.

Next **flowers** bloom among the leaves. Soon these flowers will be making seeds.

After most of the flowers are gone, big green **pods** peek out from the leaves. These pods are filled with seeds. When the pods open, the seeds will fall out.

The plant **cycle** will start all over again. These seeds grow into new plants.

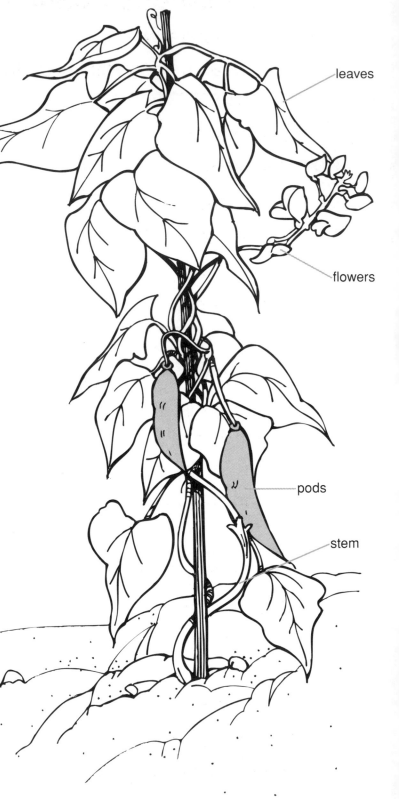

leaves

flowers

pods

stem

Read and Understand, Science • Grades 2–3 • EMC 3303

Name _____

Questions about *Plants from Seeds*

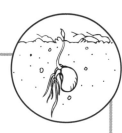

Mark the best answer.

1. What does a bean plant grow from?
   - ○ roots
   - ○ a stem
   - ○ a seed
   - ○ the ground

2. What do plants need to help them grow?
   - ○ only water
   - ○ only sunlight
   - ○ only dirt
   - ○ dirt, water, and sunlight

3. What is the job of a flower?
   - ○ to look pretty
   - ○ to smell good
   - ○ to make seeds
   - ○ to make food for the plant

4. What is the job of a green leaf?
   - ○ to look pretty
   - ○ to smell good
   - ○ to make seeds
   - ○ to make food for the plant

5. What happens first?
   - ○ the seedcase splits open
   - ○ roots start to grow
   - ○ the stem gets taller
   - ○ a shoot grows out of the ground

Name _____

Vocabulary

A. Fill in the missing words. You will not use one word.

Word Box			
cycle	seed	flowers	leaves
pods	roots	shoot	stem

1. A tiny plant is inside the _____.

2. _____ grow down from the seed.

3. A _____ grows up from the seed.

4. Green _____ make food for the plant.

5. A plant's _____ make seeds.

6. Leaves grow on the plant's _____.

7. Green _____ hold the bean plant's seeds.

B. Find **all** the words from the word box in this word search.

s	e	e	d	r	o	o	t	s
h	l	e	a	v	e	s	p	t
o	x	c	y	c	l	e	o	e
o	b	e	a	n	q	r	d	m
t	f	l	o	w	e	r	s	z

Name _____

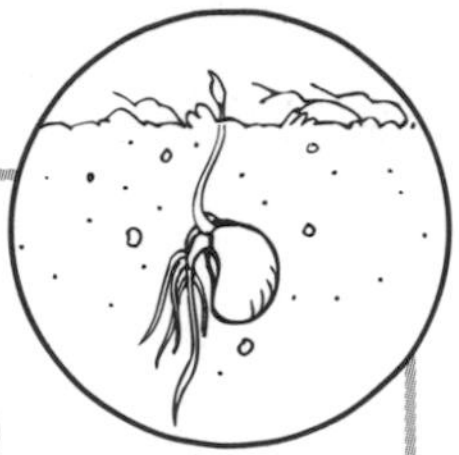

Read and Follow Directions

Put a bean seed in water. Soak it overnight. Open the seed to see what is inside. Complete the record form.

What I Used:

_____

_____

_____

What I Did:

_____

_____

_____

_____

_____

What I Saw:

A Thunderstorm

We were helping Grandpa in the garden. When we started, the sky was filled with fluffy white clouds. Then the **clouds** got bigger and darker. "I think a **thunderstorm** is on its way," said Grandpa. "Let's put the tools away and get indoors."

While we were putting the tools away, **lightning** flashed in the sky. Then we heard the rumble of **thunder**. "I think it's going to rain soon. We'd better hurry," said Grandpa.

Lightning is a giant spark of **electricity** that jumps from cloud to cloud or between a cloud and the ground.

Clouds are made up of tiny drops of water. Dark clouds have more waterdrops than white clouds. When the drops get heavy, they fall as rain.

"Look how big those black clouds are now. The **storm** is getting closer," said Grandpa. "The rain will start any minute now."

Grandpa was right. Just then **raindrops** started falling. The drops were little at first. But it wasn't long before big drops were falling. The wind started to blow. Lightning raced across the sky. Thunder rumbled with a great noise. The thunderstorm had arrived!

Grandpa got milk and cookies from the kitchen. We stood by the window with our snack and watched the thunderstorm.

Thunder is the sound that follows lightning.

1. The lightning heats the air, making it expand quickly.
2. This causes sound waves along the lightning flash.
3. We hear thunder when the sound reaches our ears.

Name _____

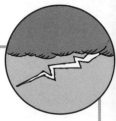

Questions about *A Thunderstorm*

Answer the questions.

1. How did Grandpa know a thunderstorm was on the way?

_____

_____

2. What is lightning?

_____

3. What is thunder?

_____

4. What happens in a thunderstorm?

_____

_____

_____

5. How are dark clouds different from white clouds?

_____

_____

Name _____

Vocabulary

Name each picture.

snack   tools   clouds

Grandpa   rain   lightning

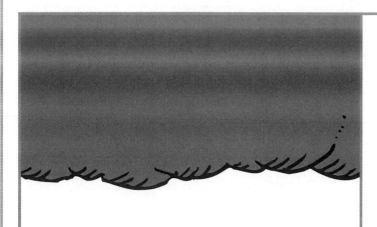

_____

_____

_____

_____

_____

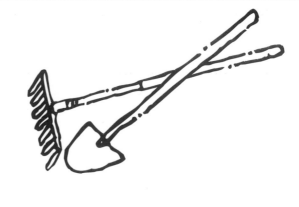

_____

Name _____

Draw and Write

Draw and write about a storm you have been in.

Scientists Use Tools

Scientists use tools to help them in their **research**. Let's watch these two scientists to see what tools they use.

Tomas

Dr. Wilson

It is Tomas's week to take care of Fuzz Ball. Fuzz Ball is the class **guinea pig**. He must give the guinea pig food and water. He needs to clean its cage. He will see if Fuzz Ball has grown.

Someone sent Dr. Wilson a strange piece of rock. He wants to look at it more closely in his **lab**.

Tomas **weighed** Fuzz Ball on the scale.

The scientist weighed the rock on a scale.

Tomas **measured** to see how long Fuzz Ball was.

He measured the rock to see how big around it was.

Fuzz Ball was chewing at his foot. Tomas took a close look to see if that foot was sore.

He cut a slice from the rock with a special saw. He looked at the slice under a **microscope**.

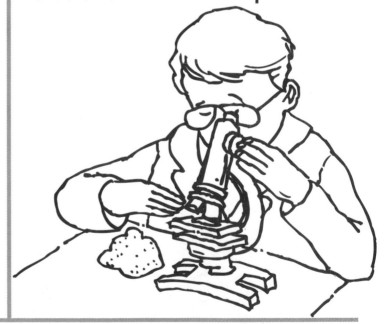

These scientists used tools to help them learn more about what they were studying. Think about how you use tools as you study new things in **science**.

Questions about *Scientists Use Tools*

Answer these questions.

1. What did the scientists in the story study?

   Tomas _____

   Dr. Wilson _____

2. How did Tomas use each of these tools?

   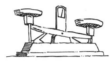 _____

     _____

   _____

3. Mark the tools Dr. Wilson used.

4. Circle the tools that make things look larger.

Vocabulary

Use these words in place of the underlined words.

slice	research	chewing
guinea pig	strange	microscope

1. Scientists use tools when they <u>study about</u> new things.

   Scientists use tools when they _____ new things.

2. Fuzz Ball was <u>biting</u> his foot.

   Fuzz Ball was _____ on his foot.

3. Dr. Wilson looked at the <u>different kind of</u> rock.

   Dr. Wilson looked at the _____ rock.

4. He cut a <u>piece</u> from the rock.

   He cut a _____ from the rock.

5. The scientist used a <u>magnifying tool</u> to make it look larger.

   The scientist used a _____ to make it look larger.

6. Tomas took care of the <u>furry little animal</u>.

   Tomas took care of the _____.

Name _____

Using Tools

1. Tomas and Dr. Wilson weighed something.
   Draw something you weighed. Why did you weigh it?

   [ ] _____

        _____

        _____

        _____

2. Tomas and Dr. Wilson measured something.
   Draw something you measured. Why did you measure it?

   [ ] _____

        _____

        _____

        _____

3. Have you ever used a magnifying glass?   yes   no
   If you answered **yes**, draw what you looked at.
   Why did you want to see it bigger?

   [ ] _____

        _____

        _____

        _____

The Moon

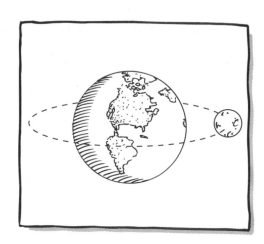

A moon **orbits** (travels around) a **planet**. Some planets don't have any moons. Other planets have many moons. **Earth** has just one moon. It is the biggest thing we see in the night sky. Sometimes you can see the Moon during the day, too.

The Moon is Earth's nearest neighbor in **space**. It is smaller than Earth. It is much smaller than the Sun. It only seems larger because it is so close to us.

The Moon has tall mountains. It has flat, dusty plains. It has large holes called **craters** (cray´ terz). Craters are made when space rocks hit the Moon.

During the day the Moon is very, very hot. During the night the Moon is very, very cold.

There is no air on the Moon. There is no water on the Moon. It is dry, dusty, and lifeless.

The Moon orbits (goes around) the Earth once a month. The Moon **rotates** (turns around) very slowly. A day on the Moon is two weeks long. It is night for the next two weeks.

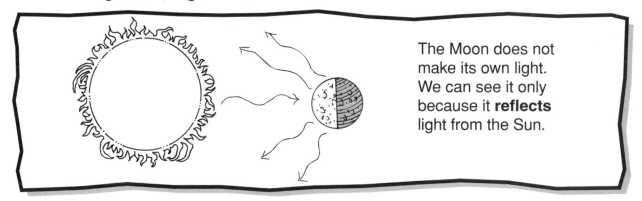

The Moon does not make its own light. We can see it only because it **reflects** light from the Sun.

If you watch the Moon for a month, it seems to change shape. The shape doesn't really change. We see different amounts of light being reflected. The shapes we see are called the Moon's **phases**. The phases follow a pattern. The pattern is repeated each month of the year.

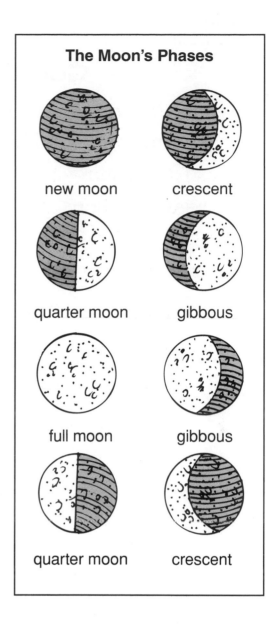

The Moon's Phases

new moon

crescent

quarter moon

gibbous

full moon

gibbous

quarter moon

crescent

Twelve men have walked on the Moon. They collected rocks to bring back to Earth. Scientists studied the rocks to learn more about the Moon.

These **astronauts** had to wear special suits. The suits kept them from getting too hot or too cold. The suits carried **oxygen** for them to breathe. The astronauts also had to carry food and water for their trip.

The Moon has changed very little for billions of years. There is no water to wear down the hills. There is no wind to move dust from place to place. The astronauts left footprints in the dust. After more than 30 years, the footprints can still be seen.

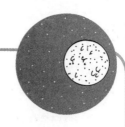

Questions about *The Moon*

Mark the best answer to the question.

1. What is the Earth's nearest neighbor in space?
   - ○ the Sun
   - ○ the Moon
   - ○ a star
   - ○ Mars

2. What will you see on the Moon?
   - ○ mountains
   - ○ craters
   - ○ dust
   - ○ all of the above

3. What does a moon orbit (travel around)?
   - ○ a planet
   - ○ a star
   - ○ a space rock
   - ○ a satellite

4. How are craters on the Moon made?
   - ○ when spaceships land on the Moon
   - ○ when astronauts dig up rocks
   - ○ when space rocks hit the Moon
   - ○ when there are storms on the Moon

5. Which of these is true about the Moon?
   - ○ The Moon makes its own light.
   - ○ The Moon reflects light from the Sun.
   - ○ The Moon is always dark.
   - ○ The Moon is a ball of fire.

Vocabulary

A. Write the correct word by its meaning.

Word Box		
astronaut	rotates	Earth
craters	moon	orbits

1. the planet we live on _____

2. the Moon turns around and around _____

3. the Moon travels around the Earth _____

4. holes in the Moon _____

5. a person who walked on the Moon _____

6. a place where nothing lives _____

B. Use words from the story to name each picture.

_____   _____   _____

Name _____

Watching the Moon

Write the names of the days in the boxes.
Look at the Moon each night for a week.
Draw what it looks like in the box on that day.

Day:	Day:	Day:

Day:	Day:	Day:

Day:	Draw the shapes of the Moon:	
	beginning	ending

How did the shape change? _____

_____

How Seeds Travel

One day, Mattie and Anne took their dog Bruno for a walk. Anne saw something growing in the front lawn. She reached down and picked it. "Mattie, look what I found. Isn't it pretty?" she asked.

"That's a **puffball**," said Mattie. "Make a wish and blow on it."

Anne blew on the puffball. She watched as little seeds flew away. "Is it true? Will I really get my wish?" she wanted to know.

"Your wish may not come true," answered Mattie. "But you just helped that plant spread its seeds. If the seeds land in **soil**, new plants will grow."

"Do other seeds travel to new places?" asked Anne.

"You are full of questions today!" laughed Mattie. "Let's go to the library. We can get a book about seeds."

Some seedpods pop open. They fling seeds into the air.

The girls and Bruno ran across a vacant lot on the corner. It was full of **weeds**. When they reached the library, Anne pointed to Bruno. She shouted, "Look at all those **stickers**."

"Each one of those stickers is a seed," Mattie told Anne. "The hooks stick to animals that pass by. Can you see how they stuck to Bruno's fur? When he rubs against something, the seeds may fall off. The seeds may grow in that new place. Now you watch Bruno while I go in and get us a book."

Mattie found a book about how seeds travel. She read the book to her little sister. She pointed out pictures of plants and their seeds.

The girls passed the vacant lot on the way home. Mattie told Anne, "Look at the vacant lot. Seeds carried here by wind or animals may grow. Then the lot will be filled with new plants."

Seeds are spread in many ways!

This is a maple seed. It has little wings. The wind blows it to new places.

Some animals eat fruit. They may spread seeds in their feces.

Squirrels and some birds bury seeds. If they forget the seeds, plants may start to grow.

Name _____

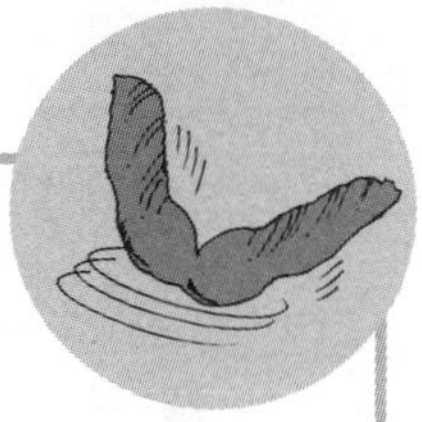

Questions about *How Seeds Travel*

1. What happened in the story?
   Draw a line from the beginning of each sentence
   to the correct ending.

Anne blew   • to Bruno's fur.

Stickers stuck   • seeds fly to new places.

When the wind blows   • a book about seeds.

At the library, Mattie got   • how seeds are spread.

The book was about   • on the puffball.

2. List four ways seeds are spread.

   _____

   _____

   _____

   _____

Name _____

Vocabulary

A. Match each word to its meaning.

1. fur   • not the same

2. different   • to get larger and older

3. grow   • animal hair

4. burst   • a place where you can borrow books

5. library   • a ball of flower seeds

6. vacant lot   • to crack open

7. puffball   • an empty piece of land in town

B. Use words from the story to name the pictures.

_____   _____   _____

Read and Understand, Science • Grades 2–3 • EMC 3303

Name _____

Spreading Seeds

Circle the part of the seed that helps it travel to new places.
Explain how the part helps the seed move.

_____

_____

_____

_____

_____

_____

_____

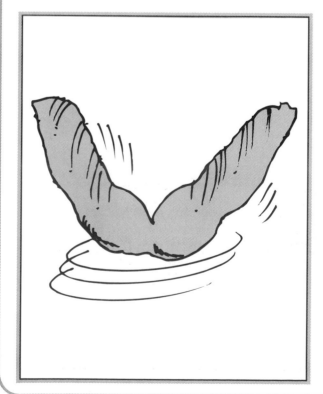

_____

_____

_____

_____

_____

_____

_____

29

Mr. de Mestral's Invention

*You never know where a new **invention** will come from.*
*This is how one **inventor** got his great idea.*

One day in 1948, **George de Mestral** and his dog went for a walk. He saw that **burrs** were sticking to his clothes and to his dog's fur. The burrs were very hard to pick off.

Mr. de Mestral wanted to find out why the burrs stuck so tightly. When he got home, he looked at the burrs with a **microscope**. He found that the burrs had sharp little hooks. The hooks clung to everything they touched. This gave him an idea! Maybe he could use what he learned from the burrs. He would make a tough new kind of **fastener**. It would not break or stick the way zippers often did.

People laughed at Mr. de Mestral when he told them his idea. But he didn't give up. He knew his idea was good. He just had to find a way to make it work.

The tiny hooks in Velcro® hold tight to little loops.

It took years of hard work. Many problems had to be solved. Finally he found a way to make tiny hooks that held tight to little loops. He was ready to sell his invention. First it needed a name. Mr. de Mestral put together two French words. He took **vel** from **velour** (velvet) and **cro** from **crochet** (hook). Together they made a great name for his new invention—**Velcro®**.

Today millions of **meters** of Velcro® are produced every year. Velcro® is used to fasten all kinds of things. Just look around you. In how many places can you see Velcro® being used? Do your shoes have Velcro® fasteners? How about your backpack or lunch bag?

Just think, all of this started because of one man's curiosity.

Questions about
Mr. de Mestral's Invention

1. How did Mr. de Mestral get his idea for Velcro®?

   _____

   _____

2. Why did the burrs stick to his clothes and the dog's fur?

   _____

3. List three ways Velcro® is used.

   _____

   _____

   _____

4. What two words were put together to name Velcro®?

   _____   _____

5. Circle the words that describe Mr. de Mestral.

   hardworking   never gave up   quitter

   lazy   problem-solver   curious

6. Look at these pictures of the burr and
   the piece of Velcro®. How are they alike?

   _____

   _____

Name _____

Vocabulary

The letters in the dark boxes below spell a mystery word. Use the clues to complete the puzzle.

1. a person who invents something

2. a measure of length

3. to hold onto something

4. to find

5. a kind of sticker with hooks

6. a tool that makes things look bigger

Word Box

burr
cling
discover
inventor
meter
microscope

What is the mystery word? _____

Using Velcro®

The story tells us ways that Velcro® is used. Go on a "Velcro® Hunt" to find other ways it is used. List the uses you find.

_____

_____

_____

_____

_____

_____

_____

_____

_____

Trees

Trees are the largest and oldest living things in the world. Trees are found in almost every **habitat** in the world. Some trees require a hot, sunny **climate**. Others live where it is cold much of the year. There are trees that are found in dry **deserts**. Others grow in wet **swamps**.

A tree is a tall plant with a stiff, woody stem. There are three main kinds of trees—**broadleaf**, **conifers**, and **palm trees**.

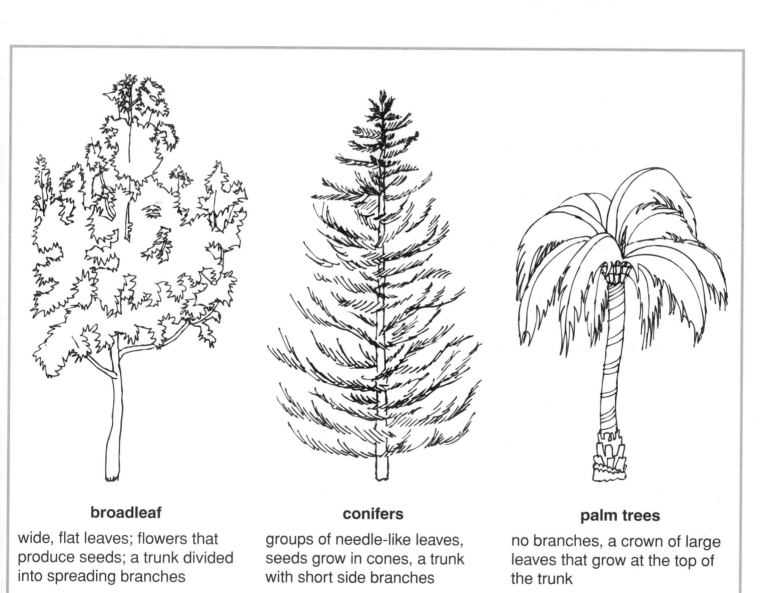

broadleaf

wide, flat leaves; flowers that produce seeds; a trunk divided into spreading branches

conifers

groups of needle-like leaves, seeds grow in cones, a trunk with short side branches

palm trees

no branches, a crown of large leaves that grow at the top of the trunk

Read and Understand, Science • Grades 2–3 • EMC 3303

Each part of the tree has its own job.

Leaves produce food for the tree. Using energy from sunlight, **chlorophyll** (the green part of leaves) makes a kind of sugar.

Branches of various sizes grow out from the trunk. The tree's leaves grow on the branches.

Bark is the cover that protects the tree from harsh weather and hungry creatures.

Seeds are kept in the cones, flowers, or fruit of the tree. The seeds are the way trees **reproduce** (make new trees). Each seed could grow into a new tree.

The thick, woody **trunk** is the stem of a tree. Tubes in the trunk carry water and food to all parts of the plant.

Roots hold the tree in the ground and **absorb** (soak up) water and minerals in the soil for the tree.

Trees are important to people. They provide lumber for our homes. They produce fruits and nuts that we eat. When leaves make food, they give off **oxygen** into the air we breathe.

Read and Understand, Science • Grades 2–3 • EMC 3303

Name _____

Questions about *Trees*

A. Mark **all** correct answers for each question.

1. Which of these is true about trees?

○ largest living things in the world
○ oldest living things in the world
○ are green all year
○ are all found in hot, dry habitats

2. What do all conifers have?

○ needle-like leaves
○ short side branches
○ colorful flowers
○ seeds in cones

3. What do all broadleaf trees have?

○ flowers that make seeds
○ branches
○ wide, flat leaves
○ seeds in cones

4. What do leaves use to make food for a tree?

○ bark
○ sunlight
○ soil
○ chlorophyll

B. Use words from the story to name these types of trees.

_____   _____   _____

37

Vocabulary

A. Match each word to its meaning.

oxygen • the flower of a tree

habitat • the green part of leaves

blossom • the gas that trees put into the air
when the leaves make food

chlorophyll • the parts of the tree that grow into
new trees

cone • the natural home of a plant or animal

roots • the tree part that holds it in the
ground; the part that absorbs water

seeds • the part of a conifer that holds
the seeds

B. Mark the correct meaning.

1. In this story, **trunk** means _____.

   ○ a kind of container in which things are packed
   ○ the nose of an elephant
   ○ the stiff, woody stem of a tree

2. In this story, **bark** means _____.

   ○ the noise a dog makes
   ○ the outside covering of a tree
   ○ to give orders in a sharp manner

Name _____

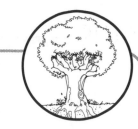

Reading a Chart

Read the chart to find the answers to the questions.

Giant Sequoias

Giant Sequoias are huge trees. They only grow in the mountains of California. They can grow to be over 300 feet tall. That is about 90 meters. They are old trees, too. Some may be as much as 4,000 years old. Many of these trees have names.

Tree	Height
General Sherman	275 feet (84 meters)
Stagg	243 feet (74 meters)
Washington	255 feet (78 meters)
Ishi Giant	248 feet (76 meters)
Robert E. Lee	255 feet (78 meters)
King Arthur	270 feet (82 meters)

1. Where do Giant Sequoias grow?
   a. in the mountains of California
   b. in the mountains of Colorado
   c. in the deserts of California

2. How old can these trees live to be?
   a. 400 years
   b. 4,000 years
   c.   40 years

3. Which two trees are the same height?
   a. Stagg and Ishi Giant
   b. General Sherman and King Arthur
   c. Washington and Robert E. Lee

Twinkle, Twinkle, Little Star

Twinkle, twinkle, little star
How I wonder what you are,
Up above the world so high
Like a diamond in the sky.
Twinkle, twinkle, little star
How I wonder what you are.

Have you ever looked up at the sky at night? Did you see stars twinkling? Have you wondered what they are?

Stars seem to be little lights in the sky. But they are not little at all. Stars seem small because they are so far away from Earth. Most stars are very big. More than a million **planets** the size of Earth could fit inside a star the size of the Sun.

Stars are balls of hot, glowing **gases**. They all give off light and heat. We see their light. We don't feel the heat they make because they are so far away. The Sun is the closest star to Earth. We can see the Sun's light. We can also feel its heat.

The Sun is the closest star to Earth.

 Read and Understand, Science • Grades 2–3 • EMC 3303

Why do stars seem to twinkle in the sky? There is a layer of moving air around the Earth. It is called the **atmosphere**. Starlight must pass through this layer before we can see it. The atmosphere makes the light bend. The movement of the light makes the stars seem to twinkle.

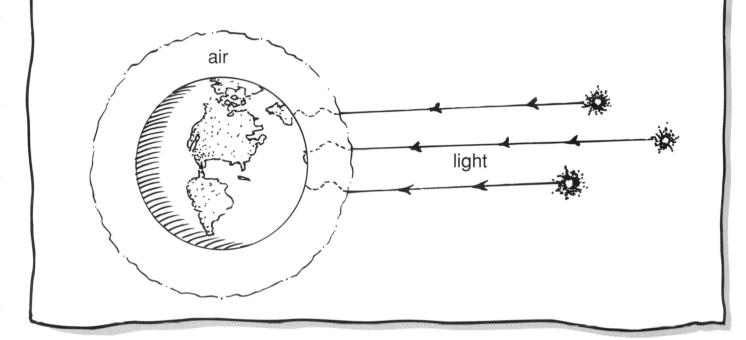

Stars shine in the sky all of the time. During the day we cannot see their light. The light from the Sun makes the sky very bright. It hides the light of the other stars. But on a clear, dark night, we can see thousands of stars in the sky.

 Read and Understand, Science • Grades 2–3 • EMC 3303

Name _____

Questions about
Twinkle, Twinkle, Little Star

Answer these questions.

1. Where does the poem tell us to look for stars?

_____

2. What is a star?

_____

3. Why do stars seem to be so small?

_____

4. Which star is the closest to Earth?

_____

5. What makes starlight seem to twinkle?

_____

_____

6. Why can't we see most stars during the day?

_____

_____

Vocabulary

A. Match each word to its meaning.

star   • a layer of air around the Earth

Earth   • a ball of hot, glowing gases

atmosphere   • to think about something

twinkle   • the star closest to the Earth

bend   • the planet we live on

Sun   • to shine with a flickering light

wonder   • to change from a straight line into a curve

B. Find the vocabulary words above in this word search.

a	t	m	o	s	p	h	e	r	e
e	s	x	s	t	a	r	s	w	b
a	p	c	o	m	e	t	u	y	e
r	a	o	c	e	a	n	n	z	n
t	c	t	w	i	n	k	l	e	d
h	e	r	e	w	o	n	d	e	r

Write a Poem

Use what you know about stars to write your own poem.
Remember, a poem can rhyme, but it does not have to.

Make a list of words that tell about stars.
Use these words to help you write your poem.

_____   _____

_____   _____

_____   _____

_____   _____

Stars

by _____

_____

_____

_____

_____

_____

_____

_____

Magnets

A **magnet** is an object that can **attract** (pull toward itself) or **repel** (push away) other objects. These objects must contain iron.

The **force** (push or pull) of a magnet is the strongest at its **poles** (the two ends). If a magnet is hanging loosely from something, the ends will point toward the Earth's poles. One end will point toward the North Pole. The other end will point toward the South Pole. A **magnetic field** surrounds the magnet. This is the space where the force of the magnet can be felt.

What will happen if you bring two magnets close together? The **opposite** poles will attract each other. That means that the south pole of one magnet will attract the north pole of the other magnet. If you put the same poles together, they will repel each other.

Everything that sticks to a magnet is metal, but not all metals stick to a magnet.

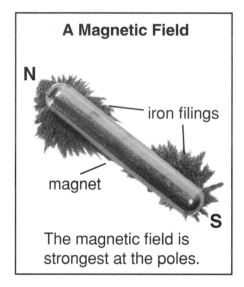

A Magnetic Field

iron filings

magnet

The magnetic field is strongest at the poles.

If you cut a magnet in half, each of the new magnets will have a north pole and a south pole. No matter how many times the magnet is cut, each piece will have both poles.

Magnets are used by businesses and in homes every day. Magnets are a part of telephones and computers. Can openers have magnets to hold the can in place. Refrigerator doors have magnets to keep them closed. Magnets are used to make doorbells ring and to turn machines such as washers off and on. Large, powerful magnets are used in junkyards to lift heavy objects.

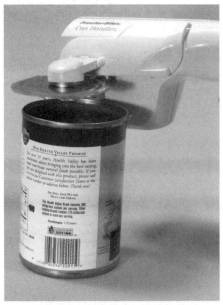

A magnet holds the can in place when you are using the can opener.

Make a Temporary Magnet

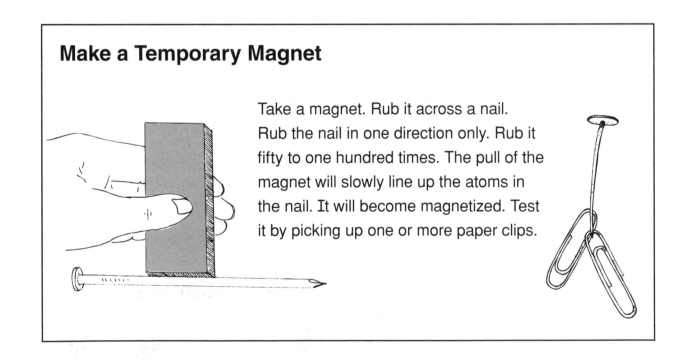

Take a magnet. Rub it across a nail. Rub the nail in one direction only. Rub it fifty to one hundred times. The pull of the magnet will slowly line up the atoms in the nail. It will become magnetized. Test it by picking up one or more paper clips.

Name _____

Questions about *Magnets*

1. What is a magnet?

_____

_____

2. Where is the force of a magnet the strongest?

_____

3. What metal is attracted to magnets? _____

4. Circle the places magnets are used.

   to hold the can in a can opener

   to peel an apple

   to keep the refrigerator door closed

   to keep the classroom door closed

   in a computer

   in a telephone

   to fasten a jacket

5. Which of these magnets will be attracted to each other?

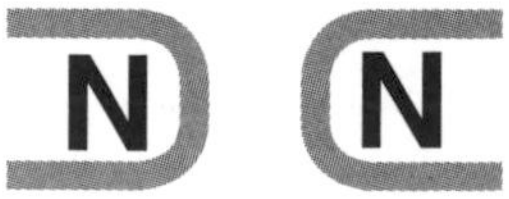

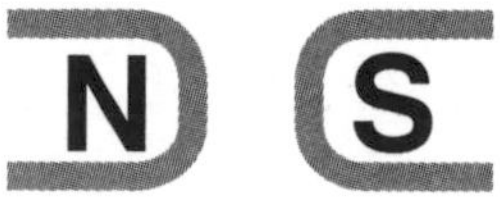

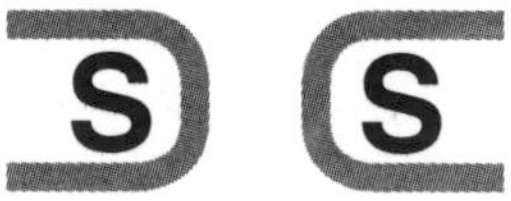

N N   N S   S S

○   ○   ○

Name _____

Vocabulary

Write each word by its meaning.

attract	magnet	repel
poles	magnetic field	iron

1. a kind of metal that is
   attracted to a magnet   _____

2. to pull toward itself   _____

3. to push away   _____

4. the two ends of a magnet   _____

5. the space around a magnet
   where the force can be felt   _____

6. an object that can attract
   or repel other objects   _____

Name _____

Note: Each student will need a copy of this page, a magnet, and a self-closing plastic bag containing a hairpin, bottle cap, brass screw, nail, paper clip, and a metal spoon.

An Experiment

Read all of the directions.
Do the experiment.
Write about what you learned.

1. Circle the metal objects that you think a magnet will attract.

2. Try to pick up each object with your magnet.

3. Draw a line under the objects that the magnet **did** pick up.
   Make an **X** on the objects that the magnet **did not** pick up.

4. What did you learn about magnets?

_____

_____

_____

_____

Moth Life Cycle

All moths change as they grow from egg to adult. In each of its four stages, the moth looks different. These changes are called **metamorphosis**.

Egg

The **female** moth lays eggs. She lays them on the leaves or stems of a plant. The leaves will become food when the eggs hatch.

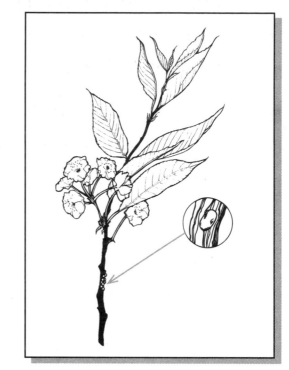

Larva

An egg hatches into a **larva**, or **caterpillar**. The hungry little caterpillar eats all of the time. As it eats, it grows. Its skin becomes too tight. A new skin grows under the old skin. The old skin splits apart and is shed. This is called **molting**. Molting happens many times as the caterpillar grows.

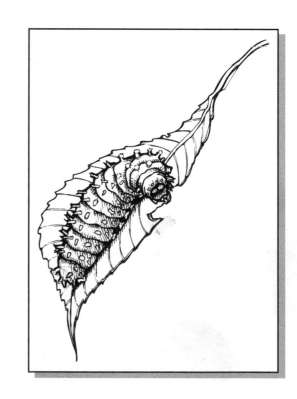

Pupa

After the caterpillar molts for the last time, it fastens itself to a twig or leaf. Then it spins a cover around its body. The cover is called a **cocoon**. This is the **pupa** stage.

Inside the cocoon, the pupa stops eating. It is beginning to change into an **adult**.

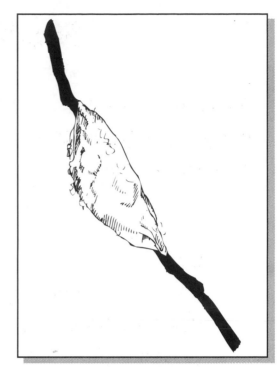

Adult

The adult moth wriggles out of its cocoon. Its wings unfold and dry. The moth is ready to fly away.

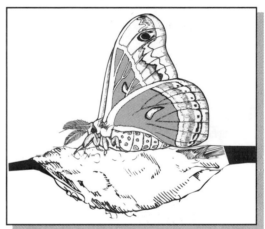

Frogs go through metamorphosis, too.

A tadpole hatches from the egg. It lives in the water. It breathes with gills. It has a tail. As the tadpole grows, lungs and legs form. Its tail disappears. At last it leaves the water as a frog.

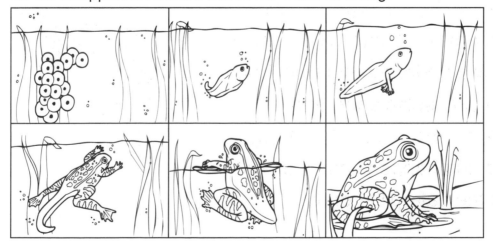

Read and Understand, Science • Grades 2–3 • EMC 3303

Name _____

What changes does a moth go through in its life cycle?

1. Cut the pictures apart.

2. Paste them in order on a sheet of writing paper.
   You will need to use both sides of the paper.

3. Write about each step.

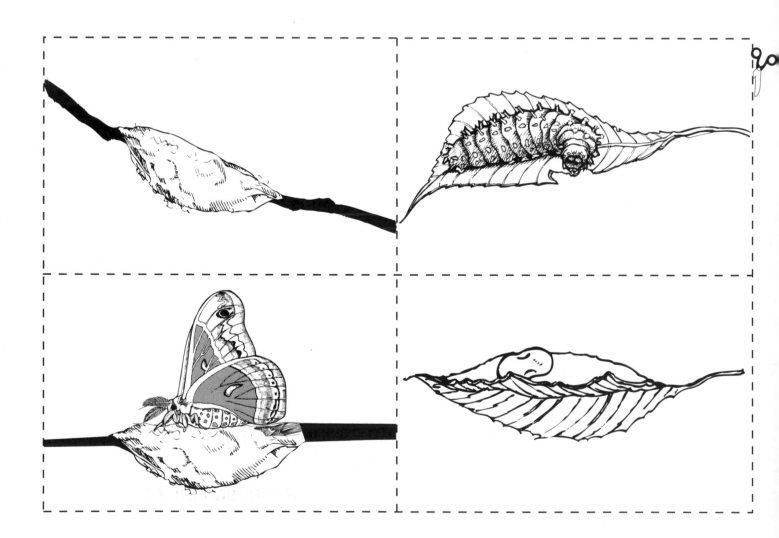

Vocabulary

A. Find words in the story to complete the sentences.

1. A moth goes through a big change called

   _____ .

2. A moth is called a _____ while it is in its cocoon.

3. Larvae are also called _____ .

4. When a caterpillar's skin gets too tight, it _____ .

B. Mark the correct meaning for the underlined word.

1. The moth will <u>fly</u> away when its wings are dry.

   ○ move through the air
   ○ a small insect
   ○ a baseball hit high in the air

2. The caterpillar eats <u>leaves</u>.

   ○ goes away
   ○ thin, flat green parts of a tree
   ○ lets alone

3. A moth makes a big <u>change</u> as it grows from egg to adult.

   ○ to become different in some way
   ○ the money you get back when you pay for something
   ○ small coins

Changes

People do not go through metamorphosis like insects and some other animals do. But you do change as you grow from a baby to an adult.

Write about some ways in which you have changed since you were a baby.

_____

_____

_____

_____

_____

_____

How will you change as you grow into an adult?

_____

_____

_____

_____

_____

Listen for the Sounds

Boom! Boom! Boom! roar the kettle drums. Rat-a-tat-tat! sing the snare drums. Listen to the sounds of drums large and small. Where does the sound come from? Why are sounds different?

When the top of the drum is struck by a stick or mallet, it **vibrates** (moves back and forth). As the drum vibrates, it moves the air around it. The sound **vibrations** travel through the air in all directions.

You hear the drum when the sound vibrations reach your ear. The vibrations are caught by your **outer ear** (the part you can see). They are carried through your **inner ear** (the part you cannot see) to your brain. Then your brain tells you that you are hearing a drum. This all happens very quickly.

> You can't see sound vibrations, but if you put cereal on the drum head and hit the drum, you can see the results of sound vibrations.

Sound can move through a **gas** such as air. It can move through a **liquid** such as water. It can move through a **solid** such as the ground we walk on.

Our world is filled with sound. Sounds such as planes or barking dogs can be unpleasant. We call these sounds noise. Sounds such as music and laughter are pleasant sounds.

Sounds can bounce off objects and make echoes.

If you are close to a jet plane, the sound will be very loud. You may need to cover your ears to protect them from the noise. If the jet is flying far away, you will hear a much softer sound. Sounds are louder close up and softer far away.

Long ago, Native Americans knew that sound travels better through the solid ground than through the air. They would put an ear to the ground to listen for the sound of moving horses or buffalo. The sound could be heard through the ground before it could be heard in the air.

Name _____

Questions about
Listen for the Sounds

A. How do you hear the sound of a drum?
Number the steps in order.

☐ The top of the drum moves back and forth (vibrates).

☐ The sound vibrations move through the air.

☐ The top of the drum is struck with a stick.

☐ Boom! You hear the drum.

☐ Your outer ear catches the sound.

☐ As the drum vibrates, it makes the air around it vibrate.

☐ Your inner ear carries the sound to your brain.

B. Pretend you are in your bedroom. Circle the loudest sound you would hear.

an alarm clock going off by your bed

a baby crying in another room

a dog barking outside in the street

Name _____

Vocabulary

A. Write each word next to its meaning.

sound	water	noise
vibration	outer ear	brain

1. something that can be heard   _____

2. the part of the ear you can see   _____

3. a quick movement back and forth   _____

4. a kind of liquid   _____

5. an unpleasant sound   _____

6. the part of the body that hears
   sound   _____

B. Use the words from the story to complete the sentences.

1. An _____ can be made when sound bounces off

   an object.

2. Sound can move through _____, _____,

   and _____.

3. Your _____ _____ carries sound vibrations to

   your brain.

Name _____

Note: Reproduce this page for each student as homework.
Students return the page on the specified date.

What Do You Hear?

Listen to the sounds around you. Make a list of the sounds you hear. Return the paper on _____.

1. Circle the pleasant sounds.
2. Make an **X** by the noisy sounds.

_____   _____

_____   _____

_____   _____

_____   _____

_____   _____

_____   _____

_____   _____

_____   _____

_____   _____

_____   _____

_____   _____

_____   _____

Phil's Science Log

Phil couldn't wait for science class to begin. He could see by the **materials** on the front table that something interesting was going to happen. There were empty soft drink bottles, balloons, and two bowls. One bowl was full of ice cubes and the other bowl was empty. "What's going to happen?" Phil wondered.

Mrs. Garcia explained that she was going to **conduct** an **experiment**. She would use the materials on the table. The students were going to watch the **demonstration**. Then they would write about what they saw.

     Read and Understand, Science • Grades 2–3 • EMC 3303

After the demonstration, Phil opened his science logbook. He began to write.

There were two empty bottles and some balloons. Alice and Jerome put the balloons over the tops of the bottles.

Mrs. G. put one bottle into the bowl of ice and one bottle into the empty bowl.

She poured hot water into the empty bowl. This is what I saw happen.

I think the hot water made the balloon blow up.

The students finished writing in their logbooks. Then they discussed what they had seen. Next, they read about the same experiment in their science books. After that, they wrote in their logbooks again. Here is what Phil wrote.

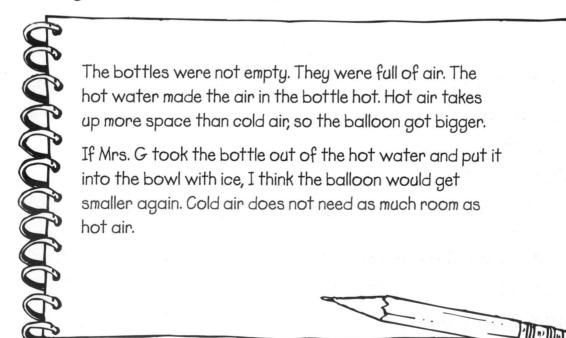

The bottles were not empty. They were full of air. The hot water made the air in the bottle hot. Hot air takes up more space than cold air, so the balloon got bigger.

If Mrs. G took the bottle out of the hot water and put it into the bowl with ice, I think the balloon would get smaller again. Cold air does not need as much room as hot air.

"Good job, Phil!" said Mrs. Garcia.

Questions about *Phil's Science Log*

A. Mark the best answer for each question.

1. Which list names all of the materials Mrs. Garcia needed for the experiment?

   ○ bowls, balloons, water, ice, jars
   ○ balloons, bowls, hot soup, bottles of water
   ○ bottles, balloons, bowls, ice cubes, hot water

2. Who helped Mrs. Garcia with the demonstration?

   ○ Phil and Alice
   ○ Jerome and Alice
   ○ Phil, Jerome, and Alice

3. Why do you think Mrs. Garcia demonstrated the experiment instead of letting her students do it?

   ○ She was in a hurry.
   ○ She did let them do the experiment.
   ○ She didn't want them using hot water.

B. Number the steps in order.

☐ Mrs. Garcia put a bottle in the bowl of ice. She put the other bottle in the empty bowl. She poured hot water into the bowl.

☐ Phil wrote in his science logbook.

☐ The class discussed what happened to the balloons.

☐ Alice and Jerome put balloons on the bottles.

☐ Phil watched what happened to the balloons.

☐ Phil wrote in his science logbook again.

Name _____

Vocabulary

A. Write each word next to its meaning.

materials   demonstration   describe

experiment   empty

1. a showing of how something is done   _____

2. to tell how someone or something looks, feels, or acts   _____

3. the objects needed to do something   _____

4. nothing inside   _____

5. a test to find out something   _____

B. Add the endings **ed** and **ing**.

explain	explain**ed**	explain**ing**
1. finish	_____	_____
2. discuss	_____	_____

observe	observ**ed**	observ**ing**
1. demonstrate	_____	_____
2. describe	_____	_____

try	tri**ed**	try**ing**
1. empty	_____	_____
2. carry	_____	_____

Name _____

A Friendly Letter

Pretend that you are Phil. Write a letter to a friend.
Explain what you learned about air from the demonstration.

_____

_____

_____

_____

_____

_____

_____

_____

_____

_____

_____

_____

_____

Rainbows

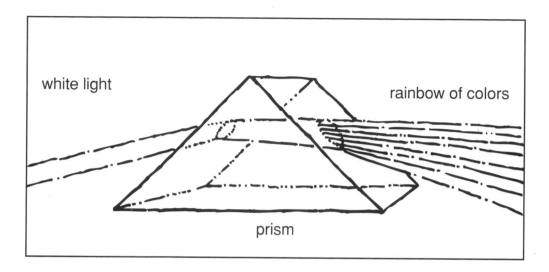

white light

rainbow of colors

prism

When the sun comes out after it has rained, you may see a **rainbow**. Where does the rainbow come from? When can a rainbow be seen?

Sunlight seems to be white. But white light is really made up of the colors of the rainbow. When sunlight passes through **raindrops**, the white light splits apart. It splits into the seven colors of the rainbow. The colors are **reflected** (bounced back) from the raindrops. Then we can see the rainbow.

Each color in sunlight bends a different amount. Red light bends the least. **Violet** light bends the most. This makes an **arc** of color bend across the sky.

You can use a **prism** to make a rainbow. Hold a prism in a beam of light. The prism breaks the light apart. You can see a rainbow on the ceiling or wall.

Another way to see white light broken into the colors of the rainbow is to blow soap bubbles. You can see the rainbow on the bubbles.

This poem can help you remember how a rainbow is made.

When rain falls down
And the sun shines behind,
You can see a rainbow.
It's easy to find.

Sunlight breaks apart
On the raindrops passing by.
Bright colors bend,
And an arc fills the sky.

Here is an easy way to remember the colors of a rainbow:

Roy G. Biv

R red

O orange

Y yellow

G green

B blue

I indigo

V violet

Name _____

Questions about *Rainbows*

Read each sentence. Mark it **true** or **false**.

1. Sunlight has all of the colors of the rainbow in it.

   ○ true   ○ false

2. The colors of the rainbow are reflected from raindrops to our eyes.

   ○ true   ○ false

3. There are 10 colors in a rainbow.

   ○ true   ○ false

4. We see a rainbow when white light breaks apart.

   ○ true   ○ false

5. Red light is on the bottom of a rainbow.

   ○ true   ○ false

6. The colors of a rainbow are in this order—red, yellow, green, blue, orange, violet, indigo.

   ○ true   ○ false

Vocabulary

A. Match each word to its meaning.

rainbow   • bits of water falling from the sky

indigo   • any part of a circle or other curved line

raindrops   • a dark shade of violet-blue

reflect   • to bounce light back from an object

arc   • an object that can separate white light into the colors of the rainbow

beam   • an arc of colors in the sky

prism   • a ray of light

B. A **compound word** is made up of two smaller words.

butter + **fly** = butterfly

Find three compound words in the story.

1. _____

2. _____

3. _____

Name _____

Read and Follow Directions

Materials

- clear jar
- water
- white paper

Steps

1. Fill the jar with water.

2. Set the jar on a sunny windowsill.

3. Put white paper beside the jar. Slide the paper around until you see a rainbow on it.

4. Draw and color the rainbow you make.

The Earth Is Always Moving

We don't feel the Earth moving, but it is never still. It is always moving in two ways.

The Earth **rotates** (spins) around. This is what causes day and night. One **rotation** of the Earth takes about 24 hours, or one day.

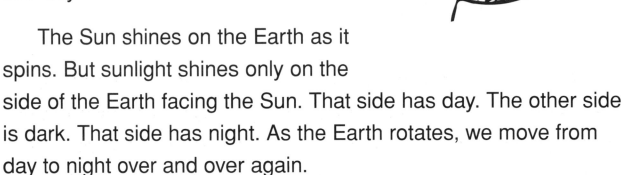

The Sun shines on the Earth as it spins. But sunlight shines only on the side of the Earth facing the Sun. That side has day. The other side is dark. That side has night. As the Earth rotates, we move from day to night over and over again.

Day and Night

Here it is night.

Here it is day.

The Earth **orbits** (travels around) the Sun. The trip takes one year, or about 365 days.

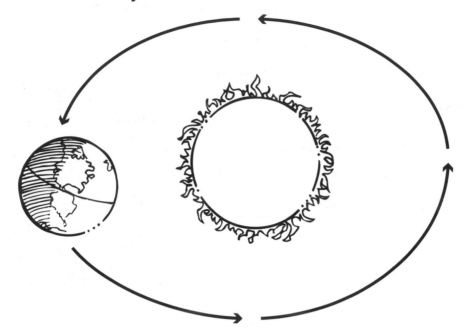

As the Earth moves around the Sun, the **seasons** change. The tilt of the Earth causes this to happen. Different parts of the Earth get more or less sunlight at different times of the year. An area gets more sunlight in the summer and less in the winter.

The Seasons

The poles are not straight up and down. The Earth is tilted.

The northern side of the Earth is tilted away from the Sun. It is winter there.

The northern side of the Earth is tilted toward the Sun. It is summer there.

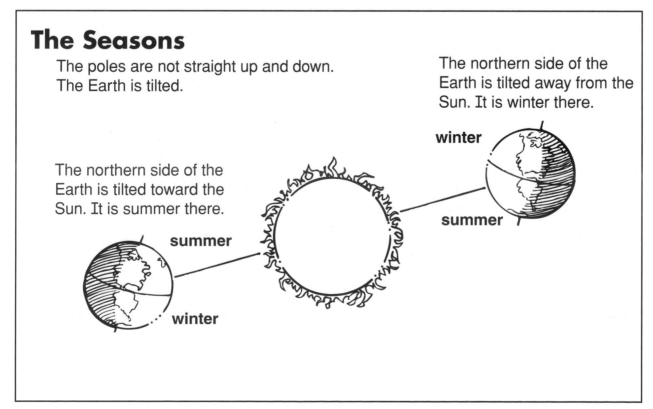

Name _____

Questions about *The Earth Is Always Moving*

A. Mark the answer.

   1. How long does it take for the Earth to rotate one time?

      ○ 1 month
      ○ 1 day
      ○ 1 week
      ○ 1 year

   2. How long does it take the Earth to travel around the Sun?

      ○ 1 month
      ○ 1 day
      ○ 1 week
      ○ 1 year

   3. What causes day and night to happen?

      ○ the Earth spinning around
      ○ the Earth traveling around the Sun
      ○ the Sun traveling around the Earth
      ○ the Moon spinning around

   4. What causes the seasons to change?

      ○ the tilt of the Earth
      ○ the amount of sunlight a place gets
      ○ the Earth moving around the Sun
      ○ all of the above

B. Name the object in space.

_____ _____ _____

Vocabulary

Use the clues to complete the puzzle.

Across

4. the Sun and the planets that travel around it

6. the time it takes Earth to travel around the Sun

Down

1. the time it takes Earth to spin around one time

2. to spin around

3. the planet we live on

5. to travel around the Sun

Word Box	
day	Earth
orbit	rotate
solar system	year

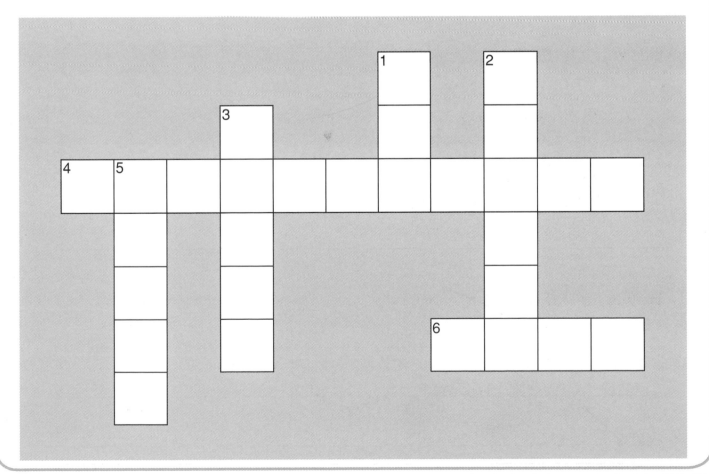

Name _____

The Seasons

As the Earth travels around the Sun, the seasons change. Pick one of the seasons. Draw and write about what it is like where you live at that time of year.

I pick:

☐ summer   ☐ autumn

☐ winter   ☐ spring

Draw:

Write:

_____

_____

_____

_____

Tanisha's Science Project

The third grade has learned many things in science this year. Now they have invited parents to come to class to see what they have learned. Tanisha is responsible for answering questions about **friction**. Her study group has made a display to show what they know. They have made a chart with facts about friction.

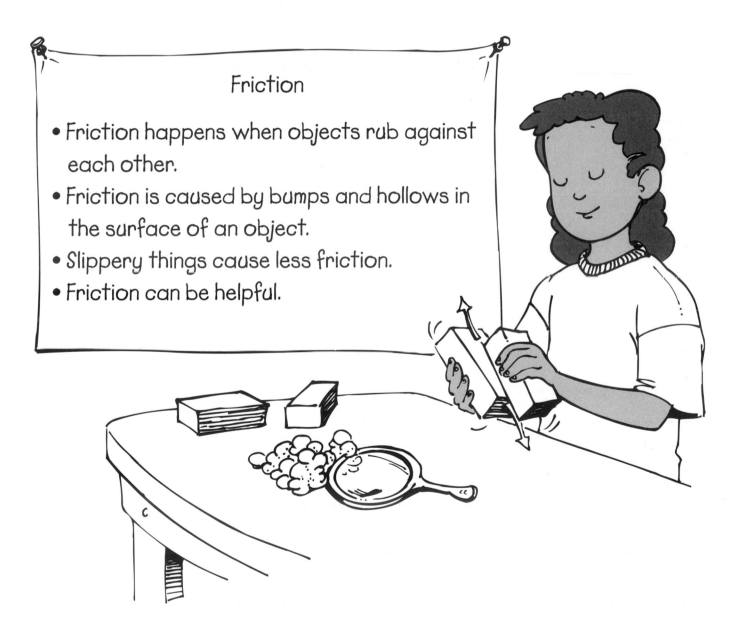

Friction

- Friction happens when objects rub against each other.
- Friction is caused by bumps and hollows in the surface of an object.
- Slippery things cause less friction.
- Friction can be helpful.

Read and Understand, Science • Grades 2–3 • EMC 3303

They prepared three experiments for people to try in order to learn about friction.

Experiment 1:

Rub a cotton ball on sandpaper.

Rub a cotton ball on a mirror.

Which was easier?

Experiment 2:

Rub two pieces of rough wood together.

Rub two pieces of smooth wood together.

Which was easier?

Experiment 3:

Rub a piece of rough wood with sandpaper.

What happens?

All week Tanisha has been thinking about answers to questions people might ask her. Now the big day is here. In a few minutes people will be coming. All of a sudden she is feeling nervous. "Am I really ready?" she wonders. She will soon find out.

A man asks, "I see your chart says that friction can be helpful. Can you explain that for me?"

Tanisha takes a deep breath and begins, "Friction helps keep the tires of your car on the road. Friction between your shoe and the ground helps when you walk or run. And you couldn't open a door without friction between your hand and the doorknob." She is ready. And she knows what she is talking about.

Friction Helps Us

Friction helps keep tires on the road.

Friction also helps to open doors.

Questions about
Tanisha's Science Project

A. Answer these questions.

1. What has Tanisha's study group made?

_____

_____

2. Why has her group prepared three experiments?

_____

_____

3. Why was Tanisha nervous?

_____

_____

4. How can friction be helpful?

_____

_____

B. Mark **all** the sentences that are true about friction.

○ Friction happens when things rub against each other.

○ Slippery things cause the most friction.

○ Bumps and hollows in surfaces cause friction.

○ Friction can be helpful.

○ Rough things cause less friction.

Name _____

Vocabulary

A. Use these words in place of the underlined words.

responsible	explain
studying	invited

1. Margo is <u>trying to learn</u> about friction.

   Margo is _____ about friction.

2. Our teacher <u>asked</u> parents to come to school.

   Our teacher _____ parents to come to school.

3. I am <u>taking care</u> of my pet dog.

   I am _____ for my pet dog.

4. Mr. Wong will <u>tell us</u> how to do the lesson.

   Mr. Wong will _____ how to do the lesson.

B. Match words that mean the opposite.

rough •   • hollows

easier   • ask

nervous   • smooth

bumps   • calm

answer   • harder

Name _____

Once When I Was Nervous

In the story, Tanisha was nervous about having to answer questions. Write about a time when you were nervous.

Why were you nervous?

_____

_____

_____

_____

How did you know that you were nervous?

_____

_____

_____

_____

How did you get over being nervous?

_____

_____

_____

_____

Building a Tree House

An old oak tree stands in a field near Danny's house. Strong, wide branches spread out from the trunk. One day as he was climbing the old oak tree, Danny had an idea. "What this tree needs is a tree house!" he shouted.

Danny and his sister gathered tools and some pieces of lumber. First they carried a ladder to the tree. This wasn't easy to do. They had to carry it across the yard, down some steps, and across the field to the tree. They leaned the ladder against the tree and hurried back to the garage. By the time they had carried a piece of lumber to the tree, the children were tired. "We're never going to get everything we need out to the tree," complained Danny.

"Don't give up, Danny," said Jan. "Let's think. Maybe we can come up with a better idea."

As the children sat thinking, Uncle Fred came along. "Hey, kids. What are you up to?" he called. Danny explained that they wanted to build a tree house in the oak tree.

"It's too hard to get everything we need out to the tree," explained Jan.

Uncle Fred grinned. "I think I can help." He explained that there were **simple machines** that could make their work easier.

Uncle Fred found a **wheelbarrow** in back of the garage. The children filled it with tools and laid pieces of lumber across the top. They pushed the wheelbarrow across to the steps. "How are we going to get the wheelbarrow down these steps?" asked Jan.

wheel

Uncle Fred told them not to worry. He went back into the garage. In a few minutes he returned with a piece of plywood and a rope. He leaned the wide board against the steps and helped push the full wheelbarrow down the **ramp**. "Why do you need that rope?" asked Danny.

ramp

"Wait and see," said Uncle Fred. "Let's get this stuff out to the tree. I'm anxious to get started on this tree house."

pulley

When they reached the tree, Uncle Fred climbed up the ladder and threw the rope over a branch. Soon he had made a simple **pulley** and was lifting the tools and lumber up into the tree.

"You make it look so easy, Uncle Fred," Jan said admiringly.

"It is easy when you know the right machines to use," answered Uncle Fred. "The wheels, the ramp, and the pulley all helped make our work easier."

By the end of the week, Uncle Fred, Jan, and Danny were sitting inside the finished tree house having a picnic to celebrate a job well-done.

Questions about
Building a Tree House

A. Answer these questions.

1. What was Danny's great idea?

_____

2. What problem did the children have in building the tree house?

_____

3. How did simple machines help with their problem?

_____

B. Mark the answer to the questions.

1. Which simple machine was used to carry the lumber to the tree?

   ○ wheel and axle (wheelbarrow)
   ○ pulley
   ○ none of the above

2. Which simple machine was used to lift the lumber and tools up into the tree?

   ○ inclined plane (ramp)
   ○ pulley
   ○ none of the above

3. Which simple machine was used to get the wheelbarrow down the steps?

   ○ inclined plane (ramp)
   ○ wheel and axle (wheelbarrow)
   ○ pulley
   ○ none of the above

 Read and Understand, Science • Grades 2–3 • EMC 3303

Vocabulary

A. Fill in the missing word.

complained	ramp	lumber
celebrate	tree house	simple machines
explained	idea	picnic

1. Uncle Fred made a _____ on top of the steps.

2. "This is too hard," _____ Danny.

3. _____ _____ make work easier.

4. Danny's great _____ was to build a _____

   _____.

5. "I can help solve your problem," _____ Uncle Fred.

6. They had a _____ to _____ finishing the tree house.

7. They used _____ to build the floor and walls.

B. Match the words that have about the same meaning.

eager • • tired

worn out • • lifting

collected • • anxious

began • • grinned

raising • • gathered

smiled • • started

Name _____

Simple Machines

How did Uncle Fred use these simple machines?

pulley

inclined plane   **wheel and axle**

Write the kind of simple machine Uncle Fred used.
Then explain how the simple machine made the work easier.

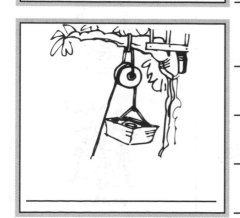

_____

_____

_____

_____

_____

_____

_____

_____

_____

_____

_____

_____

Read and Understand, Science • Grades 2–3 • EMC 3303

Honeybees

cells in honeycomb

Have you ever heard the saying "busy as a bee?" There is one type of honeybee that really is always busy. From the time she crawls out of her **cell** in the **hive**, a worker bee works!

Thousands of bees live together in a hive. There is a **queen bee** that has one big job. She lays the eggs that will hatch into new young bees. There are a few male bees called **drones**. Their only job is to mate with the queen. The rest of the bees in a hive are **worker bees**. These little insects are well named. They do everything else.

For the first 20 days of life, worker bees do jobs inside the hive. These **"house bees"** clean and repair the wax cells of the hive. They feed the growing young bees. They turn **nectar** (sweet juice from flowers) into honey. They guard the hive from intruders.

After 20 days, a worker bee is ready to be a **forager.** She takes short flights to learn her way around, and then she begins her trips to collect water, **pollen**, and nectar. A hardworking forager may make 10 trips a day. She may visit 100 flowers on each trip.

queen

drone

worker

Read and Understand, Science • Grades 2–3 • EMC 3303

The sweet smell of flowers helps the forager find them. She lands on a petal and sips the nectar. She carries it to the hive in a special honey stomach. The nectar is made into honey.

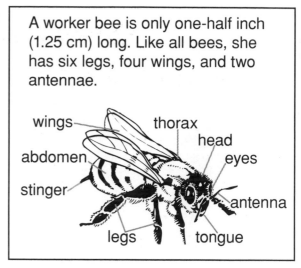

A worker bee is only one-half inch (1.25 cm) long. Like all bees, she has six legs, four wings, and two antennae.

wings, thorax, head, eyes, antenna, tongue, legs, stinger, abdomen

The forager also collects flower dust **(pollen)**. The yellow pollen sticks to hairs on her body. She brushes the pollen onto her hind legs to carry it back to the hive. Pollen and honey are food for the bees.

When a worker bee finds a good place to gather nectar, she can tell other bees how to get to the same place. She does this by dancing. The way she moves in the dance tells other bees which way to go. There is the **round dance** that tells how to get to places close to the hive. There is the **waggle dance** that tells how to get to places far from the hive.

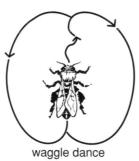
waggle dance

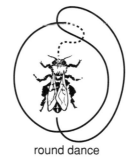
round dance

A worker bee has a short life—only a few weeks. But in that time she does important work for her hive. And her hard work helps us, too. As bees gather nectar, they carry pollen from flower to flower. Plants need the pollen to make seeds. Without bees, many fruits and flowers that we use would disappear.

 Read and Understand, Science • Grades 2–3 • EMC 3303

Questions about *Honeybees*

1. What are the three kinds of honeybees in a hive?

   _____   _____

   _____

2. What is the job of the queen bee?

   _____

3. What kind of honeybee is a drone? What is the job of the drone?

   _____

   _____

4. When does a "house bee" become a "forager"?

   _____

   _____

5. List five jobs worker bees do for the hive.

   _____

   _____

   _____

   _____

   _____

Vocabulary

Read about the parts of a worker bee.

- Her long **tongue** works like a straw. She uses it to sip nectar from flowers.

- She feels and smells with two **antennae** on her head.

- She has four **wings** that are used for flying.

- She crawls and climbs about on her six **legs**.

- She has three body parts—a **head** in front, a **thorax** in the middle, and an **abdomen** in back.

- She uses a **stinger** at the end of her abdomen to defend herself.

Label the parts of this bee.

Name _____

Reading a Table of Contents

A **table of contents** shows what is in a book. Read this table of contents about honeybees to find the answers to the questions.

Table of Contents

1. How many chapters are in this book?

   _____

2. On what page can you read about worker bees?

   _____

3. What will you find if you turn to page 14?

   _____

4. List two things you can learn when you read a table of contents.

   _____

   _____

Animals Without a Backbone

Think about animals for a minute.

Did you think about your pet dog or cat? Did you think about an elephant in the zoo? These animals are all alike in one important way. They all have a **skeleton**. Part of that skeleton is the **backbone**.

Most animals in the world don't have backbones. Spiders, insects, worms, and snails don't have backbones. Many kinds of animals in the sea don't have backbones. Some of these animals have a hard outside covering. It is like an **external** skeleton. Some of these animals have no skeleton at all.

Let's look at two common animals without backbones.

Spiders

Every kind of spider is alike in these ways. They have a hard outside covering. They have eight legs. Most spiders have eight eyes, too. They have fangs to poison their prey. They have two main body parts. Their heads and chests are joined in front. They have an **abdomen** in back.

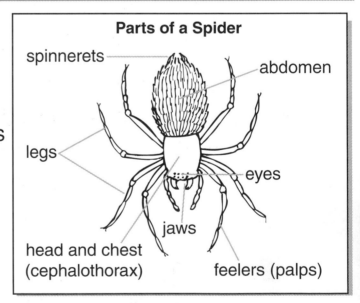

Parts of a Spider

spinnerets

abdomen

legs

eyes

jaws

head and chest (cephalothorax)

feelers (palps)

Some people are afraid of spiders. But most spiders don't harm people. In fact, they are very helpful. Spiders eat harmful insects.

Spiders spin silk. They make the silk in **spinnerets**. They pull the silk out using their back legs. The light, strong silk is a sticky liquid. It hardens in the air. Spiders spin silk to make webs or other traps. Silk is also used to make a case around a spider's eggs.

Insects

There are thousands of kinds of insects. They live all over the world. Insects are different sizes, shapes, and colors. But they are all alike in several ways.

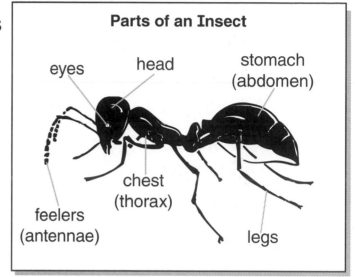

Parts of an Insect

eyes   head   stomach (abdomen)

chest (thorax)

feelers (antennae)

legs

All insects have six legs. They have two **antennae** on their heads. An insect's body is divided into three parts. Most kinds of insects have two or four wings.

There are insects that fly, hop, crawl, dig, and swim. Some insects eat plants. Some insects eat other animals. There are even insects that eat feathers and fur. Has a mosquito ever bitten you? It was after a dinner of your blood!

The next time you are outdoors, look around. See how many different kinds of animals without a backbone you can find.

Questions about *Animals Without a Backbone*

A. Mark the correct answer.

1. Which of these animals has a backbone?

   ○ elephant
   ○ insect
   ○ spider
   ○ none of the above

2. Which of these animals do not have a backbone?

   ○ insects
   ○ spiders
   ○ worms
   ○ all of the above

3. Which sentence is true?

   ○ Most animals in the world have a backbone.
   ○ Most animals in the world do not have a backbone.
   ○ There are no animals with a backbone.

B. Draw a spider and an insect. Show their parts.

Name _____

Vocabulary

A. Match each word to its meaning.

skeleton   • dangerous

external   • the part of the spider that makes silk

harmful   • the main bone down the back

spinnerets   • outside

insect   • the bones of the body

backbone   • a small six-legged animal

B. Mark **all** the correct answers. Use what you learned about insects and spiders to help you.

1. Mark the insects.

2. Mark the spiders.

     93   Read and Understand, Science • Grades 2–3 • EMC 3303

Name _____

Comparing Insects and Spiders

How are an insect and a spider alike? How are they different?
Fill in the spaces to show your answers. Use information from the
story to help you.

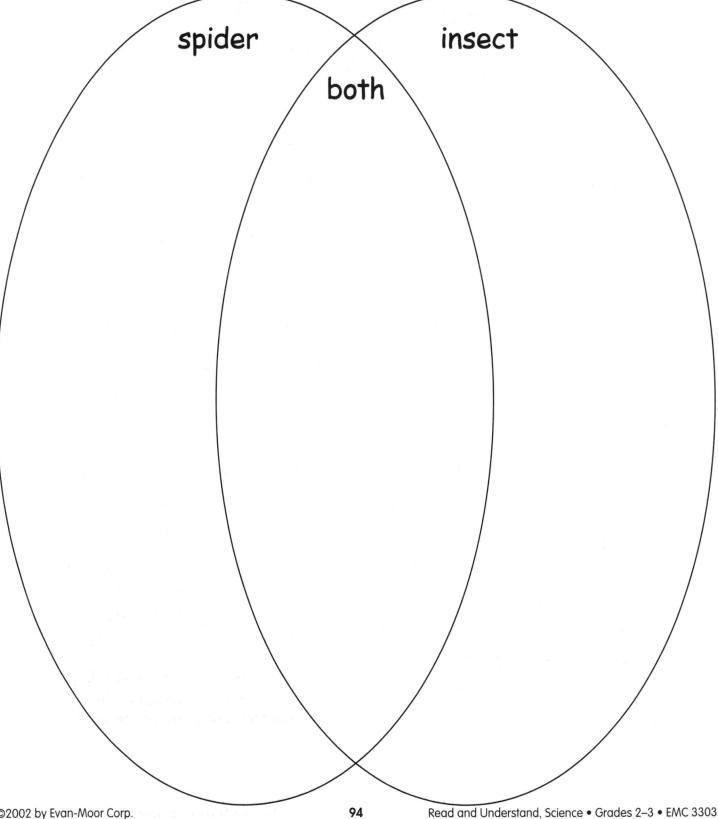

spider   both   insect

Take a Closer Look

Scientists base their ideas on what they **observe**. This means they need to take a close look at what they are studying. Some scientists might need to look at objects far off in space. Other scientists might look at tiny objects here on Earth.

Since early times, scientists have been curious about the **universe**. They could see the Moon and thousands of stars. But they wanted to know more than just what their eyes could show them.

Then the **telescope** was invented. How excited those early scientists must have been. At last they had a tool to help them take a closer look. Now they could see the closest planets. They could see the rings around **Saturn**. They could see that **Jupiter** had many moons. But they wanted to know more.

The largest telescopes are in **observatories** around the world.

Read and Understand, Science • Grades 2–3 • EMC 3303

As time passed, more powerful telescopes were made. It is now possible to see objects more clearly. Today's powerful telescopes allow scientists to look millions of miles out in space.

Telescopes help scientists take a closer look at distant objects. But scientists also need to look at very small things. The invention of the **microscope** made this possible. It let scientists look at details too small for the human eye to see. They could see things as small as blood **cells**. They could examine tiny parts of insects and plants.

Today, a more powerful microscope can make something look thousands of times bigger than it is. That is taking a really close look!

When you use a magnifying glass, you are using a simple microscope.

Name _____

Questions about *Take a Closer Look*

A. Mark the correct answer to each question.

1. Which invention helps scientists take a closer look at objects in space?

   ○ microscope
   ○ telescope
   ○ laser
   ○ magnifying glass

2. Which invention helps scientists take a closer look at small objects here on Earth?

   ○ microscope
   ○ telescope
   ○ laser
   ○ Hubble space telescope

3. Why do scientists want to take a closer look at things?

   ○ to learn more about stars and planets
   ○ to learn more about small objects on Earth
   ○ to learn more about the parts of our bodies
   ○ all of the above

B. Write the name of each tool.

      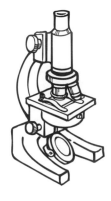

_____   _____   _____

     Read and Understand, Science • Grades 2–3 • EMC 3303

Name _____

Vocabulary

A. Write each word by its meaning.

curious	cells	observe
details	universe	telescope

1. to look at carefully   _____

2. eager to know   _____

3. everything there is; all things   _____

4. small parts of living matter   _____

5. small bits of information   _____

6. a tool for looking at objects
   that are far away   _____

B. Match words that mean the same.

small   • examine

look   • distant

close   • near

far   • tiny

Name _____

Note: Students will need a magnifying glass to complete this page.

Take a Closer Look

Find a small object. It can be something like a leaf, an insect, or a rock. Look at it using just your eyes. Record what you see.

Get a magnifying glass. Look at the object again using the magnifying glass. Record what you see.

I looked at a _____.

This is what it looked like using just my eyes.

This is what it looked like using a magnifying glass.

Keeping Warm

Heat always flows from something warmer to something cooler. If a warm object touches a cooler object, heat will flow from the warm object to the cool object. When you feel something hot or cold, you are actually feeling the flow of heat.

If you hold an ice cube, your hand is warmer than the ice cube. The heat flows from your hand to the cold ice. Your hand feels colder because you are losing heat. If you hold a cup of hot chocolate, the heat moves from the warm cup to your cool hand. Your hand feels warmer because you are getting heat.

During cooking, heat from the stove moves from the hot burner to the cold food. The flow of heat warms the food. This movement of heat is called **conduction**.

Heat moves toward something cooler.

Heat doesn't travel through all materials in the same way. It moves very easily through things made of metal. Metal is a good **conductor** of heat. That is why metal pans are used for cooking. Poor conductors of heat are called **insulators**. Wood and plastic are insulators. That is why cooking pans often have wooden or plastic handles.

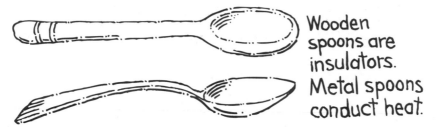

Wooden spoons are insulators. Metal spoons conduct heat.

If you go out on a cold day and are not dressed warmly, you will feel cold. Heat will flow from all parts of you into the cold air and ground. You can stop some of that heat flow by wearing warm clothes. Your clothing acts as an insulator to keep you warm. Several layers of clothes are warmer than one. Each layer traps a layer of insulating air.

Insulation is used to keep homes warm. It keeps heat from escaping from the warm house to the cold outdoors.

Insulation can also keep cold things cold. Refrigerators and freezers have layers of insulation in their walls. This keeps cold air trapped inside and keeps warm air outside.

Name _____

Questions about *Keeping Warm*

Answer the questions.

1. What makes something feel hot or cold?

   _____

   _____

2. In what direction does heat flow?

   a. from warm to warm

   b. from cool to warm

   c. from warm to cool

3. Which of these is the best conductor of heat?

   a. plastic

   b. metal

   c. wood

4. Why do you feel cold when you go outside on a cold day?

   _____

   _____

5. How can you keep warm when you go outside on a cold day?

   _____

   _____

6. Why is it important to insulate a house?

   _____

   _____

Vocabulary

A. Match each word to its meaning.

flow

• something used to keep heat from flowing out or in

object

• the movement of heat

conduction

• movement

conductor

• a thing

insulation

• something that allows heat to flow through it easily

B. Use the words above to complete these sentences.

1. Metal is a good _____ of heat.

2. My refrigerator has _____ to keep heat out.

 Read and Understand, Science • Grades 2–3 • EMC 3303

Name _____

Which Way Will the Heat Flow?

Look at each picture carefully. Draw arrows to show which direction the heat will flow. Then write a sentence telling why it is flowing in that direction.

Heat goes out from the hot flame to the cooler air.

It's Not Just Dirt!

Think of it!
Tiny **organisms** in the soil called **bacteria** help to rot the dead plants and animals.

Reach down and pick up a handful of dirt. You are holding something very important—soil. **Soil** covers much of the land on Earth. This soil is not just dirt. If you look very closely, you will find that soil contains many interesting surprises.

Soil contains small pieces broken off larger rocks. These are mixed with **organic material** (decaying plants and animals). Air and water are also a part of soil. They fill the spaces between pieces of rock and **decayed matter**. It takes many years to make good soil.

Plant roots and tunneling creatures such as ants and worms turn and move the soil around. This mixes the parts of the soil together. It also leaves holes and tunnels where water and air can collect.

Read and Understand, Science • Grades 2–3 • EMC 3303

Different kinds and colors of soil are found in different places around Earth. Each type of soil contains rocks in different sizes. **Sandy soil** has the largest bits of rock. If you rub a handful of sandy soil it will feel rough. **Silt** has smaller bits of rock. When rubbed, it feels like flour. **Clay** has the smallest bits of rock. It feels sticky and is hard to squeeze. Clay is also the heaviest type of soil. Soil often contains a mixture of types of soil in different amounts. This is called **combination soil**.

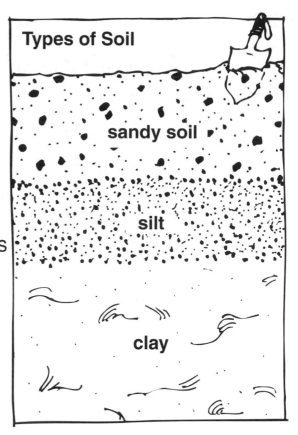

There are layers of soil on Earth. **Topsoil** (the very top layer) is the best soil. This is where most plants grow. The **subsoil** under the topsoil is made up mostly of broken rocks that reach down to **solid** rock.

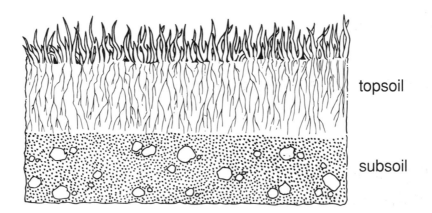

Soil is one of Earth's most important **resources**. Most of the plants on Earth grow in soil. The soil holds plants in place so they are not washed or blown away. Plants get water, some food, and minerals from soil. Humans and animals need soil, too. We eat the plants that grow in the soil.

Name _____

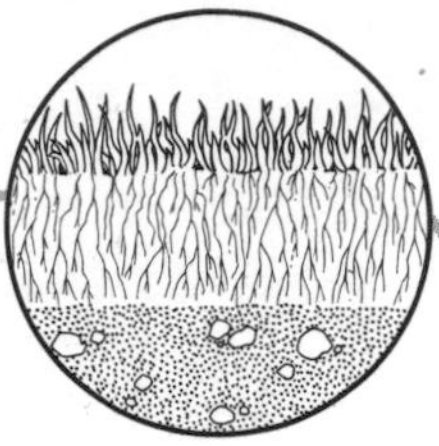

Questions about *It's Not Just Dirt!*

A. Mark the best answer.

1. What is in soil?

   ○ pieces of rock
   ○ decayed plants and animals
   ○ air and water
   ○ all of the above

2. What is organic material?

   ○ water
   ○ rocks
   ○ living things
   ○ all of the above

3. Where do most plants grow?

   ○ in the subsoil
   ○ in the topsoil
   ○ on a farm
   ○ none of the above

4. Which of these are types of soil?

   ○ clay
   ○ silt
   ○ combination
   ○ all of the above

B. Why is soil so important to living things? Give 3 reasons.

_____

_____

_____

Name _____

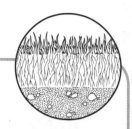

Vocabulary

Use these words in place of the underlined words.

silt	resource	creatures
topsoil	organic	soil

1. <u>Dirt</u> covers much of the Earth.

   _____ covers much of the Earth.

2. <u>Soil with small bits of rock</u> feels smooth and powdery like flour.

   _____ feels smooth and powdery like flour.

3. Soil is an important <u>material in nature that fills our needs</u>.

   Soil is an important _____.

4. Most plants grow in the <u>upper layer of soil</u>.

   Most plants grow in the _____.

5. <u>Decaying plants and animals</u> are part of soil.

   _____ materials are part of soil.

6. Small animals <u>such as worms and ants</u> tunnel in the soil.

   Small _____ tunnel in the soil.

Name _____

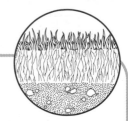

Will It Become Soil?

Think about what you learned in the story to help do this.

Read these words. Write them in the correct place.

apple core	foil wrapper
marble	potato peelings
plastic spoon	paper
leaves	nails
tin can	chicken bone

These things can decompose and will become part of the soil.

_____

_____

_____

_____

_____

_____

These things will not decompose. They will not become part of the soil.

_____

_____

_____

_____

_____

_____

Now, think of at least two other objects of each kind to write below.

_____   _____

_____   _____

Platypus

The platypus is a strange **mammal** found only in Australia. It has fur like other mammals. It is **warm-blooded** like other mammals. A baby platypus drinks its mother's milk like other mammal babies. But a platypus is different from other mammals in an important way. A female platypus lays eggs.

What Does a Platypus Look Like?

An adult platypus is about 24 inches (62 **centimeters**) long. It weighs about 4 pounds (1.8 **kilograms**). Its thick brown fur is waterproof. It has fur everywhere except on its feet and bill. Its short legs have webbed feet. It has a wide, flat tail. Sharp nails on its feet are used to dig **burrows**.

The strangest part of a platypus's body is its bill. The bill looks like a duck's bill. It is really a **snout**. The snout is covered in rubbery skin.

A male platypus has **spurs** on its ankles. The spurs hold poison that can kill small animals.

How Does a Platypus Move?

A platypus spends most of its time in freshwater ponds and streams. It has **adapted** to its water **habitat**. It can close its eyes, **nostrils**, and ears when it swims under water.

A platypus is a great swimmer. It uses its front feet to move through the water. It steers with its back feet and flat tail. On land, a platypus is awkward.

How Does a Platypus Gather Food?

A platypus does most of its hunting at night. It eats tiny animals like crayfish, snails, shrimp, and worms. It uses its sensitive rubbery snout to scoop up tiny **prey** from the muddy water at the bottom of the pond or stream. The platypus then stores the food in its cheek pouches, returning to the surface to eat. It does not have teeth. It grinds its food on rough **ridges** inside its cheeks.

A platypus stores fat in its tail. This fat is used for energy when it cannot find food.

How Are Platypus Babies Born?

A female platypus builds her nest in a tunnel. She digs the tunnel in the riverbank. She lines the nest with grass and leaves. She lays two or three small eggs in the nest. The eggs have leathery shells. She holds the eggs between her tail and belly for about two weeks. This keeps the eggs warm.

After they hatch, the tiny babies crawl up on the mother's belly. They drink milk that comes from tiny openings in her **belly**.

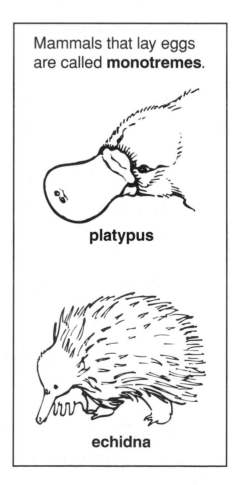

Mammals that lay eggs are called **monotremes**.

platypus

echidna

Name _____

Questions about *Platypus*

Answer these questions.

1. Where in the world are platypuses found?

_____

2. What is a monotreme?

_____

3. Describe how a platypus gets its food.

_____

_____

_____

_____

4. In what ways has a platypus adapted to life in the water?

_____

_____

_____

5. Label the parts of this platypus.

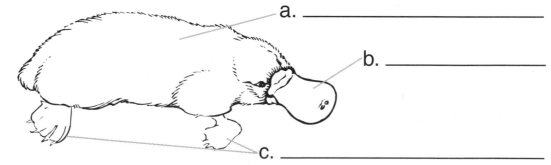

a. _____

b. _____

c. _____

6. How is a platypus like other mammals?

_____

_____

Name _____

Vocabulary

Read the sentences. Fill in the missing words.

cheek pouches	poison	snout
monotreme	adapted	spurs
mammal	tunnel	riverbank

1. A _____ baby drinks milk from its mother's body.

2. The platypus has _____ to a water habitat.

3. A mammal that lays eggs is called a _____.

4. A platypus stores food in its _____ until it comes to the water's surface.

5. The leathery bill of a platypus is not really a bill. It is

   a _____.

6. A male platypus has _____ on his ankles. They

   hold _____ that can kill small animals.

7. A female platypus builds her nest at the end of a _____

   that she digs in the _____.

Name _____

Platypus or Duck?

Mark the facts that are true about a duck, a platypus, or both of these animals.

	Platypus	Both	Duck
1. lives in the water most of the time			
2. is covered in feathers			
3. is covered in fur			
4. lays eggs			
5. has a hard bill			
6. has a snout that looks like a leathery bill			
7. has webbed feet			
8. can swim			
9. can fly			
10. feeds milk to its young			

         Read and Understand, Science • Grades 2–3 • EMC 3303

What Happened to My Pizza?
The Story of Digestion

My body needs **energy** to do its job. I get my energy from the food I eat. But the food must be changed before my body can use it. Here's the amazing story of how this happens.

My **digestive system** is the part of my body that changes the food. Part of it works like a food blender, chopping, mashing, and mixing the food until it is in very tiny pieces. Follow this piece of pizza through my body to see what happens along the way.

Just looking at the pizza and smelling it makes my mouth begin to water. I can't wait to take a big bite.

I chew the pizza in my mouth. Chewing breaks the food into smaller pieces. **Saliva** (juices) in my mouth mixes with the pizza and makes it softer.

Read and Understand, Science • Grades 2–3 • EMC 3303

I swallow, and the pizza goes down my **esophagus** (food tube) and into my bag-like **stomach**. In my stomach, muscles squeeze and mash the pizza. It is mixed with stomach juices to break it into smaller and smaller pieces.

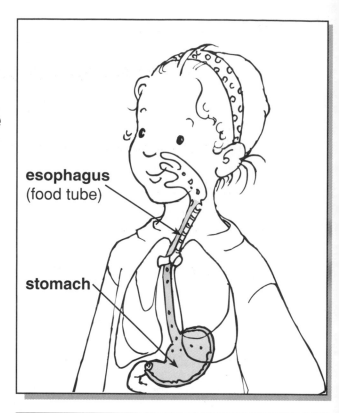

Next the food moves to my **small intestine**. Here more juices are added to the pizza to help digest it. The tiny pieces of pizza are broken into even smaller pieces called **molecules**. The molecules are small enough to go through the walls of the small intestine into my **bloodstream**. My blood carries the molecules from the digested pizza all around my body.

But some food is not used in my body. This part of the food is packed together in my **large intestine**. When I go to the bathroom, the **feces** are pushed out of my body.

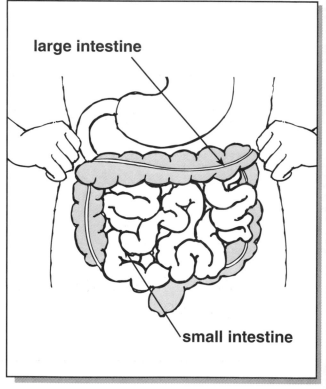

My body uses the digested food for energy. It gives me the energy I need to grow and learn. It gives me the energy I need to work and play.

Name _____

Questions about *What Happened to My Pizza?*

1. Where does your body get energy to work?

   _____

   _____

2. How does food get from your mouth to your stomach?

   _____

   _____

3. What happens to the food your body cannot use?

   _____

   _____

4. Why does food have to be changed before your body can use it?

   _____

   _____

   _____

5. Why is it important to eat a good breakfast before you go to school?

   _____

   _____

   _____

Vocabulary

A **glossary** gives the meanings of words from the story or book being read. Read the words and meanings in the glossary below. Use a word from the glossary to complete each sentence.

Glossary	
digestive system	the part of a body that changes food so the body can use it for energy
esophagus	a tube that carries food from the mouth to the stomach
feces	food not used by the body; removed from the body when you go to the bathroom
saliva	watery liquid in the mouth that helps make food easier to swallow
stomach	a large bag of muscle that holds food and helps digest it before it goes on to the small intestine

1. The food my body can't use is called _____.

2. When I swallow food, it goes down my _____

   into my _____.

3. My _____ changes food into energy for my body.

What Happened to My Pizza?

Number the steps in the correct order to show how food is digested. Then write the number of each step on the picture to show where that part of digestion happens.

☐ Food goes down my esophagus (food tube) to my stomach.

☐ The liquid food goes through my small intestine. Here tiny molecules of food go into my blood.

☐ The food my body doesn't use is pushed out of my body as feces.

☐ My teeth chew the food to make little pieces. The food mixes with saliva.

☐ My stomach juices make the food into a thick liquid.

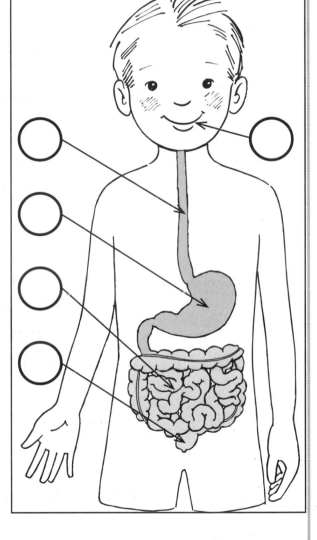

A North American Desert

A desert looks like a place with few living things. But take a closer look. There may be hundreds of kinds of plants and animals living there.

What Is a Desert?

A desert is any place that gets less than 10 inches (25 centimeters) of rain each year. Deserts can be hot or cold. **Death Valley** is very hot. It is a desert. **Siberia** is very cold. It is a desert, too.

The deserts in North America are hot deserts. They get very little rainfall. A lot of that rain quickly **evaporates** (goes back into the air).

Desert Plants

Plants growing in the desert must live for a long time without rainfall **(drought)**. They have had to learn to **adapt** (to live in) the heat.

Some deserts are covered in sand. The wind blows the sand into mounds or hills called **dunes**.

Some deserts are covered with rocky cliffs and hills. Blowing winds carve rocks into strange shapes.

Other deserts are flat, dry plains of soil and gravel.

Many plants store water when it rains. Some, such as cactus, have thick stems that store water. Others store the water in their leaves. The leaves may be covered with a layer of wax or little hairs. They may be very small. All of these adaptations keep the water from evaporating. Desert plants can live for many months using the stored water.

 Read and Understand, Science • Grades 2–3 • EMC 3303

Many desert plants have deep roots. These roots can pull water from far below the surface. Others have surface roots that spread over a large area. They collect water on the ground when rain does fall.

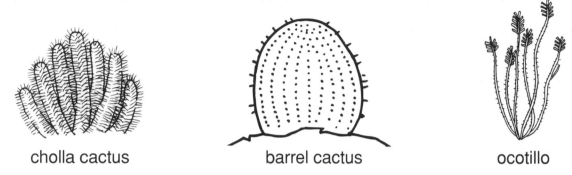

cholla cactus   barrel cactus   ocotillo

The seeds of small flowering plants lie in the soil until it rains. Then they sprout, grow, flower, and make new seeds. All of this has to happen quickly. Soon it will be too hot and dry for the plants to live.

Desert Animals

It isn't easy for animals living in a desert. These animals must be able to live with very little water. Some desert animals get all the water they need from the plants or seeds they eat. Some meat-eaters **(predators)** get water from the prey they eat. Other animals get water by licking the **dew** that collects on plants and rocks during the night.

Heat is also a problem. Some animals rest in shady spots during the day. Others live underground. They only come out at night. Some desert animals have hard scales that keep them from drying out in the hot sun.

Name _____

Questions about
A North American Desert

Answer these questions.

1. What is a desert?

_____

_____

_____

2. How do desert plants survive without much rainfall? Give at least three examples.

_____

_____

_____

3. How do desert animals survive without much rainfall? Give at least three examples.

_____

_____

_____

Vocabulary

Complete the crossword puzzle.

Across

2. to become used to; to adjust to conditions

6. to turn into a vapor

7. the opposite of cold

8. a place getting less than 10″ (25 cm) of rain each year

Down

1. a meat-eating animal

3. a long time without rain

4. moisture from the air that collects in drops on cool surfaces during the night

5. a desert plant with a thick stem that usually has spines

Word Box

- adapt
- cactus
- desert
- dew
- drought
- evaporate
- hot
- predator

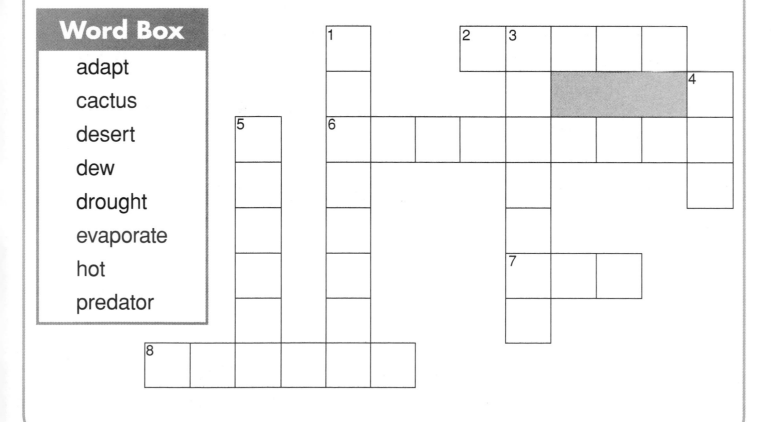

Habitats

Compare a desert habitat with the place where you live.

_____

_____

_____

_____

How are they the same?

_____

_____

_____

_____

How are they different?

_____

_____

_____

_____

Turn this paper over and draw a picture of your habitat.

Why Recycle?

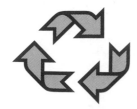 As we work and play, we are left with piles of trash. Not everything needs to end up in the garbage can. Many things can be recycled.

What Is Recycling?

Recycling is making something new from **materials** that we have already used. Used glass can be made into new bottles. Empty cans can be made into new cans. Used paper can be made into new paper. Plastic trash can be made into new useful forms.

Why Should We Recycle?

Recycling reduces the amount of trash we throw away. There is less to put into **landfills** and **incinerators**. One half of the trash we throw out could be recycled.

Recycling a 6-foot stack of newspaper saves a 35-foot tall tree!

Recycling saves **resources** by using the same materials over again. This saves **energy**. It takes less energy to turn recycled materials into new products than to make something brand new.

Recycling creates less **pollution**. It helps keep our air and water clean.

How Can We Recycle?

Communities have different rules for recycling. Some communities ask that everything except yard waste go into one container. Others ask that items be separated. You might be asked to separate newspapers from magazines and other kinds of paper. You might need to tie cardboard into bundles. Cans, glass, and plastic might go into three different containers.

Here are some ways to prepare trash for the recycling bins:

- Wash glass and remove any metal lids. Do not put mirrors or broken glass in the recycle bin.

- Separate newspapers from other kinds of paper.

- Rinse cans before putting them into the recycling bin.

- Rinse plastic bottles and remove the lids and caps. Flatten the bottles so they don't take up so much room in the recycling bin.

Questions about *Why Recycle?*

A. The story asked and then answered three questions.
Read the questions below. Answer them in your own words,
using what you learned from the story.

1. What is recycling?

_____

_____

2. Why should we recycle?

_____

_____

3. How can we recycle?

_____

_____

B. Explain how your family recycles.

_____

_____

_____

_____

_____

Name _____

Vocabulary

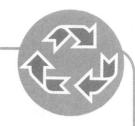

Write each word by its meaning. You will not use all of the words.

Word Box			
separate	container	products	pollution
recycle	trash	yard wastes	plastic

1. grass clippings and weeds _____

2. to divide into groups; keep apart _____

3. what harmful things in the environment cause _____

4. anything useless and thrown away _____

5. to process something so it can be used again _____

6. a box, can, jar, or carton used to hold something _____

7. objects that are made or grown _____

Name _____

Let's Recycle

Make a poster to encourage people to recycle.
Use words and pictures to get your message across.

Plastic
A Manmade Product

Take a close look around your classroom. It is filled with furniture, books, paper, and crayons. The list goes on and on. All of these are **manmade** products. These products may be made of wood, metal, plastic, or a combination of materials. Let's take a closer look at a manmade material—plastic.

Plastic is one of the most useful manmade materials. It is lightweight. It doesn't break very easily. It is inexpensive to make. It can be formed into many kinds of products.

Scientists have invented many kinds of plastics. The method used to form the plastic depends on the shape that is needed.

Molded Plastic

Plastic **pellets** are melted. The **liquid** plastic is poured into specially shaped molds. The plastic is pressed into the **mold**. When it has cooled and hardened, the object will have the same shape as the mold.

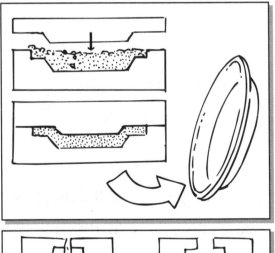

Blown and Molded Plastic

Melted plastic is poured into a mold. Air is blown into the mold. When it is cooled and removed, the object is hollow inside. This is how useful objects such as bottles are made.

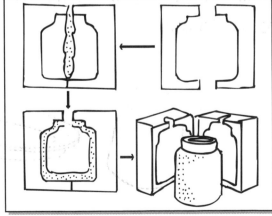

Read and Understand, Science • Grades 2–3 • EMC 3303

Spun Plastic

Liquid plastic is spun into threads. When the threads are cool, they are woven into plastic rope or cloth.

Rolled Plastic

Liquid plastic is cooled a bit. While it's still soft, the plastic is moved through rollers. The rollers press the plastic into long, flat sheets. These are made into products such as shopping bags and plastic wrap.

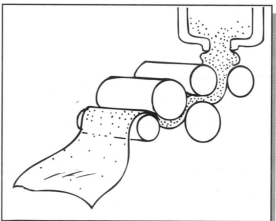

Plastic is used to make packing pellets, **insulation**, and a kind of **solid foam**. The solid foam is used in making some kinds of furniture.

Another great thing about plastic is that it can be recycled. Old plastic can be ground up, melted, and formed into new products for us to use.

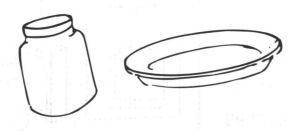

Name _____

Questions about *Plastic*

Answer these questions.

1. Why is plastic so useful? Give four examples.

    _____

    _____

    _____

    _____

2. What method is used to make each of these objects?

    plate _____   plastic wrap _____

    bottle _____

3. How is plastic recycled?

    _____

    _____

    _____

4. List four things you use that are made of plastic.

    _____   _____

    _____   _____

Name _____

Vocabulary

Complete the crossword puzzle.

Across

1. an object that is made or grown
4. not very heavy
6. what a thing is made from
8. doesn't cost much

Down

2. a mixture
3. empty inside
5. a hollow shape in which something is formed
7. drawn out and twisted into thread

Word Box

combination	hollow
inexpensive	lightweight
material	mold
product	spun

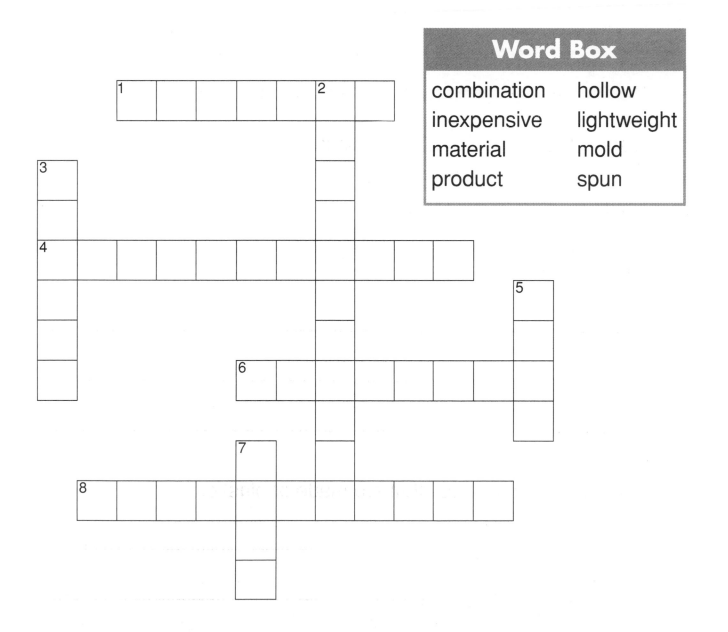

From Nature or Manmade?

Some things we use come from nature. The apple you eat at lunch comes from nature. Some things we use are manmade. The pencil you write with is manmade.

Go on a search to find other examples of each kind. List them here.

From Nature	Manmade

Fossils

With each new **fossil** find, we learn more about **ancient** life on Earth. But what are fossils? And how are they formed?

Fossils are hardened remains of animals or plants that lived millions of years ago. They lived long before there were any people on our planet. Scientists study these fossils to find out about ancient life. Fossils of both land and sea life have been found.

T-Rex Found in South Dakota Desert

In 1990 fossil hunter Sue Hendrickson made a big discovery in the South Dakota desert. She found the fossil of a Tyrannosaurus Rex. It is the largest, most complete skeleton found so far.

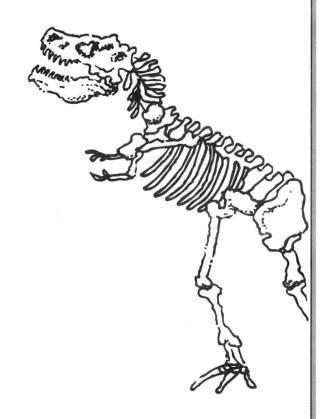

Scientists estimate it would have stood 13 feet (4 meters) tall at the hip. It would have been 41 feet (12 1/2 meters) long. The fossil even showed the dinosaur's last meal. It was duck-billed dinosaur! The fossil appears to be 67 million years old. The fossil is on permanent display at Chicago's Field Museum.

After an animal or plant died, its soft parts **decayed**. The hard parts remained lying on the ground. Some of the hard parts were covered with sand, dirt, or mud. Over millions of years, **minerals** seeped in. They took the place of the bones or other parts. What was left was a fossil in that same shape.

How does it become a fossil?

The form decomposes. It is covered with dirt or mud.

The largest fossils are of the dinosaurs. There are fossils of whole skeletons. There are fossils of nests of dinosaur eggs. Teeth, footprints, and prints of skin have all been found.

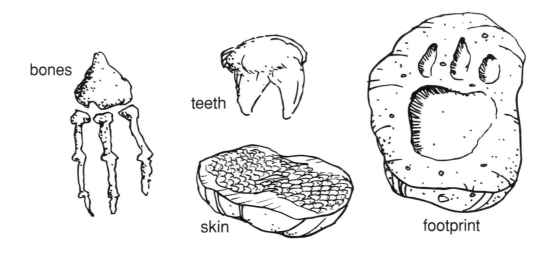

bones

teeth

skin

footprint

Scientists called **paleontologists** study fossils to learn about **prehistoric** life. Paleontologists don't spend all of their time studying fossils. They go on field trips to collect fossils. They study the rocks where fossils are found. Each new fossil discovery adds to our understanding of what Earth was like in prehistoric times.

Name _____

Questions about *Fossils*

A. Answer the questions.

1. What was special about the T-Rex in the story?

    _____

    _____

2. Why is it important to study fossils?

    _____

    _____

3. Fossil bones have been found. What other kinds of dinosaur
   fossils have been discovered? Name four.

    _____   _____

    _____   _____

B. Number the steps in order to show how a dinosaur became
   a fossil.

   [ ] Sand, dirt, or mud covered the dinosaur bones.

   [ ] The fossilized bones were found by a fossil hunter.

   [ ] The dinosaur decayed, leaving only its bones.

   [ ] Minerals seeped into the bones. They hardened into
   rock that looked just like the bones.

   [ ] A dinosaur died.

Vocabulary

Write each word after its meaning.

fossil	T-Rex	decay
minerals	prehistoric	paleontologist
skeleton	dinosaur	ancient

1. a short name for Tyrannosaurus Rex _____

2. the time before histories were written _____

3. the bones of a body _____

4. the hardened remains of a plant or
   animal that lived millions of years ago _____

5. to rot; decompose _____

6. substances dug out of the earth that
   are not a plant or an animal _____

7. a kind of reptile that lived millions of
   years ago _____

8. a scientist who studies fossils to find out
   about life on Earth millions of years ago _____

9. belonging to a time long, long ago _____

138

Name _____

A Fossil Discovery

Imagine that an exciting new fossil has been found. You are the newspaper reporter given the job of writing about the find. Answer these questions about the find, then write up your article. Add an illustration of the fossil.

What was found? _____

Who found it? _____

Where was it found? _____

When was it found? _____

Why is it an important discovery? _____

_____
headline

_____

_____

_____

_____

_____

_____

_____

_____

_____

Answer Key

Page 7
1. a seed
2. dirt, water, and sunlight
3. to make seeds
4. to make food for the plant
5. the seedcase splits open

Page 8
A. 1. seed   5. flowers
   2. roots   6. stem
   3. shoot   7. pods
   4. leaves

B.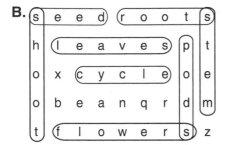

Page 9
Answers will vary, but could include the following:
What I Used:
   water, seed, paper towel
What I Did:
   I put the seed in water.
   I let it stay in the water all day.
   I opened the seed.
   I looked inside.
What I Saw:
   Drawing should show the inside of a seed.

Page 12
1. The clouds were getting bigger and darker.
2. It is a giant spark of electricity in the sky.
3. It is the sound that comes from lightning.
4. Answers will vary, but could include the following:
   Raindrops fall.
   Wind blows.
   Lightning flashes.
   Thunder rumbles.
5. Dark clouds have more waterdrops.

Page 13
clouds   snack
Grandpa   lightning
rain   tools

Page 14
Answers will vary.

Page 17
1. Tomas—Fuzz Ball or guinea pig
   Dr. Wilson—a strange rock
2. scale—to weigh Fuzz Ball
   measuring tape—to find out how long Fuzz Ball was
   magnifying glass—to look at Fuzz Ball's sore foot
3. These should be marked:
   measuring tape, scale, microscope
4. These should be circled:
   magnifying glass, microscope

Page 18
1. research   4. slice
2. chewing   5. microscope
3. strange   6. guinea pig

Page 19
Answers will vary.

Page 22
1. the Moon
2. all of the above
3. a planet
4. when space rocks hit the Moon
5. The Moon reflects light from the Sun.

Page 23
A. 1. Earth   4. craters
   2. rotates   5. astronaut
   3. orbits   6. Moon

B. crater, the Moon, Earth

Page 24
Answers will vary.

Page 27
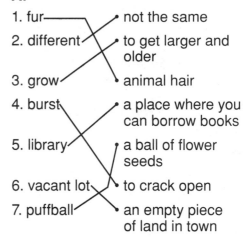

1. Anne blew ————— to Bruno's fur.
   Stickers stuck ——— seeds fly to new places.
   When the wind blows — a book about seeds.
   At the library, Mattie got — how seeds are spread.
   The book was about — on the puffball.

2. Answers will vary, but could include the following:
   stick to animal's fur
   blow in the wind
   seedpods fling seeds
   animals eat fruit and drop seeds in feces

Page 28
A.
1. fur ——————— not the same
2. different ——— to get larger and older
3. grow —————— animal hair
4. burst ————— a place where you can borrow books
5. library ———— a ball of flower seeds
6. vacant lot ——— to crack open
7. puffball ———— an empty piece of land in town

B. puffball, lawn, stickers

Page 29
burr   The hooks hold onto things the seed touches.

maple seed When the wind blows, the wings carry the seed through the air.

Page 32
1. He saw burrs sticking to his clothes and his dog's fur.
2. The burrs had sharp little hooks.
3. Answers will vary.
4. velour, crochet
5. These words should be circled:
   hardworking, never gave up, problem-solver, curious
6. They both have hooks that can stick to things.

Page 33
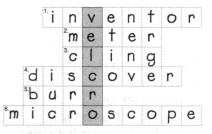

The mystery word is velcro.

Page 34
Answers will vary.

Page 37
A. 1. largest living things in the
world
oldest living things in
the world
2. needle-like leaves
short side branches
seeds in cones
3. flowers that make seeds
branches
wide, flat leaves
4. sunlight, chlorophyll

B. palm, broadleaf, conifer

Page 38
A.
oxygen — the flower of a tree
habitat — the green part of
leaves
blossom — the gas that trees
put into the air when
the leaves make
food
chlorophyll — the parts of the tree
that grow into new
trees
cone — the natural home of
a plant or animal
roots — the tree part that
holds it in the
ground; the part that
absorbs water
seeds — the part of a conifer
that holds the seeds

B. 1. the stiff, woody stem of a tree
2. the outside covering of a tree

Page 39
1. a
2. b
3. c

Page 42
1. Up above the world so high
2. A star is a ball of hot,
glowing gases.
3. They are so far away.
4. The Sun is our closest star.
5. The light bends when it passes
through the atmosphere.

6. The Sun is so bright that it hides
the light of the other stars.

Page 43
A.
star — a layer of air around
the Earth
Earth — a ball of hot, glowing
gases
atmosphere — to think about
something
twinkle — the star closest to
the Earth
bend — the planet we live on
Sun — to shine with a
flickering light
wonder — to change from a
straight line into a
curve

B.

a	t	m	o	s	p	h	e	r	e
e	s	x	s	t	a	r	s	w	b
a	p	c	o	m	e	t	u	y	e
r	a	o	c	e	a	n	n	z	n
t	c	t	w	i	n	k	l	e	d
h	e	r	e	w	o	n	d	e	r

Page 44
Answers will vary.

Page 47
1. A magnet is something that can
attract or repel other objects.
2. It is strongest at the poles.
3. Iron is attracted to magnets.
4. These should be circled:
to hold the can in a can opener,
to keep the refrigerator door
closed, in a computer, in a
telephone
5. N S

Page 48
1. iron   4. poles
2. attract   5. magnetic field
3. repel   6. magnet

Page 49
1. Answers will vary.
3. A line should be drawn under
the bottle cap, nail, paper clip,
and metal spoon. There should
be an X on the hairpin and
the brass screw.

4. Answers will vary, but should
explain that not all metals are
attracted to a magnet.

Page 52
1. An egg is laid on a leaf.
2. The egg hatches and
the caterpillar eats leaves and
gets bigger.
3. The caterpillar spins a cocoon.
4. The adult moth comes out of the
cocoon.

Page 53
A. 1. metamorphosis 3. caterpillars
2. pupa   4. molts

B. 1. move through the air
2. thin, flat green parts of a tree
3. to become different in
some way

Page 54
Answers will vary.

Page 57
A. 2
4
1
7
5
3
6

B. an alarm clock going off by
your bed

Page 58
A. 1. sound   4. water
2. outer ear   5. noise
3. vibration   6. brain

B. 1. echo
2. gases, liquids, solids
3. inner ear

Page 59
Answers will vary.

Page 62
A. 1. bottles, balloons, bowls, ice
cubes, hot water
2. Jerome and Alice
3. She didn't want them using
hot water.

B. 2
4
5
1
3
6

Page 63

A. 1. demonstration   4. empty
2. describe   5. experiment
3. materials

B. 1. finish**ed**, finish**ing**
2. discuss**ed**, discuss**ing**
1. demonstrat**ed**, demonstrat**ing**
2. describ**ed**, describ**ing**
1. empti**ed**, empty**ing**
2. carri**ed**, carry**ing**

Page 64
Answers will vary.

Page 67
1. true   4. true
2. true   5. false
3. false   6. false

Page 68
A.

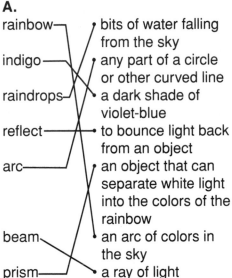

rainbow — an arc of colors in the sky
indigo — a dark shade of violet-blue
raindrops — bits of water falling from the sky
reflect — to bounce light back from an object
arc — any part of a circle or other curved line
beam — a ray of light
prism — an object that can separate white light into the colors of the rainbow

B. rainbow, sunlight, raindrops

Page 69
Drawings will vary, but the colors of the rainbow should be correct.

Page 72
A. 1. 1 day
2. 1 year
3. the Earth spinning around
4. all of the above

B   Moon, Sun, Earth

Page 73

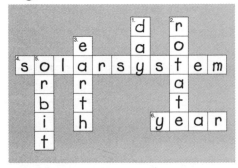

Page 74
Answers will vary.

Page 77
A. 1. They made a display about friction including a chart and three experiments.
2. The experiments will help people learn about friction.
3. She wasn't sure she was ready to answer questions people might ask.
4. Answers will vary, but could include the following:
Friction helps keep a car's tires on the road.
Friction between your shoe and the ground helps when you walk.
Friction helps you turn a doorknob.

B. Friction happens when things rub against each other.
Bumps and hollows in surfaces cause friction.
Friction can be helpful.

Page 78
A. 1. studying   3. responsible
2. invited   4. explain

B.

rough — smooth
easier — harder
nervous — calm
bumps — hollows
answer — ask

Page 79
Answers will vary.

Page 82
A. 1. His idea was to build a tree house.
2. It was hard to get everything they needed out to the tree.
3. Simple machines made the work easier.

B. 1. wheel and axle (wheelbarrow)
2. pulley
3. inclined plane (ramp)

Page 83
A. 1. ramp
2. complained
3. simple machines
4. idea, tree house
5. explained
6. picnic, celebrate
7. lumber

B.

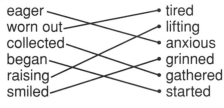

eager — anxious
worn out — tired
collected — gathered
began — started
raising — lifting
smiled — grinned

Page 84
ramp—The wheelbarrow could go down without bumping over the steps.

wheelbarrow—A lot of material could be carried at one time. The wheels made it easier to move things.

pulley—They could lift things up into the tree without carrying them up the ladder. It helped them lift heavy things.

Page 87
1. queen, drone, worker
2. She lays eggs that will hatch into new bees.
3. A drone is a male bee. Its job is to mate with the queen.
4. after 20 days
5. Answers will vary, but could include the following:
clean and repair wax cells
feed young bees
make honey
guard the hive
collect water, pollen, and nectar

Page 88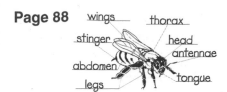

Page 89
1. seven
2. page 7
3. how honey is made
4. Any two of the following:
   what is in the book
   what page each chapter
   begins on
   how long each chapter is

Page 92
A. 1. elephant
   2. all of the above
   3. Most animals in the world
      do not have a backbone.

B. Drawings will vary, but must
   show the correct parts

Page 93
A.
skeleton— dangerous
external— the part of the
spider that
makes silk
harmful— the main bone
down the back
spinnerets— outside
insect— the bones of the
body
backbone— a small six-legged
animal

B. 1. The two insects should
be circled (ladybug and fly.)
2. The two spiders should be
circled.

Page 94
Answers will vary, but should
include the following:
spider—8 legs; most have eyes;
2 main body parts
both—no backbone; a hard
outside covering
insect—6 legs; 2 antennae; 3 body
parts; most have 2 or 4 wings

Page 97
A. 1. telescope
   2. microscope
   3. all of the above

B. telescope, magnifying glass,
   microscope

Page 98
A. 1. observe   4. cells
   2. curious   5. details
   3. universe   6. telescope
B.
small— examine
look— distant
close— near
far— tiny

Page 99
Answers will vary.

Page 102
1. When heat flows from something
   hot to something cold.
2. c
3. b
4. You feel cold because heat flows
   from your warm body to the cold
   air and ground.
5. If you wear warm clothing, it
   helps keep your heat from
   flowing away.
6. Insulation keeps heat from
   escaping from the warm house
   to the cold outside.

Page 103
A.
flow— something used to
keep heat from
flowing out or in
object— the movement
of heat
conduction— movement
conductor— a thing
insulation— something that
allows heat to flow
through it easily

B. 1. conductor   2. insulation

Page 104
1. Heat flows from the warm
   body to the cold air.
2. Heat flows from the hot
   chocolate to the cooler air.
3. Heat flows from the warmer
   air to the cold ice cubes.
4. Heat flows from the hot oven
   to the cold cookie dough.

Page 107
A. 1. all of the above
   2. living things
   3. in the topsoil
   4. all of the above

B. Answers will vary, but could
   include the following:
   Most plants grow in soil.
   Soil holds plants in place.
   Plants get water, some food, and
      minerals from soil.
   Humans and animals eat plants
      that grow in the soil.

Page 108
1. Soil   4. topsoil
2. silt   5. Organic
3. resource   6. creatures

Page 109
can   **won't**
decompose   **decompose**
apple core   marble
leaves   plastic spoon
potato peelings   tin can
paper   foil wrapper
chicken bone   nails

Page 112
1. Platypuses are found in
   Australia.
2. A monotreme is a mammal
   that lays eggs.
3. A platypus scoops up small
   prey from the muddy water
   using its snout.
4. It can close its eyes, ears,
   and nostrils under water, and
   is a good swimmer using
   webbed feet.
5. a. body or fur
   b. snout
   c. webbed feet
6. It has fur on its body. It is
   warm-blooded. The babies drink
   milk from the mother's body.

Page 113
1. mammal   6. spurs, poison
2. adapted   7. tunnel, riverbank
3. monotreme
4. cheek pouches
5. snout

Page 114

1. both
2. duck
3. platypus
4. both
5. duck
6. platypus
7. both
8. both
9. duck
10. platypus

Page 117

1. My body gets energy from the food I eat.
2. The food goes from my mouth down my esophagus to my stomach.
3. It is packed together in my large intestines. It is pushed out of my body when I go to the bathroom.
4. When I eat, the pieces of food are too big to be used by my body. The food has to be broken into very tiny pieces.
5. A good breakfast gives you energy to learn and grow and to work and play.

Page 118

1. feces
2. esophagus, stomach
3. digestive system

Page 119

2
4
5
1
3

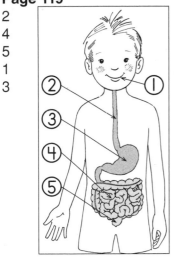

Page 122

1. A desert is a place that gets very little rainfall each year. A desert can be hot or cold.

2. Answer could include the following:
   Some plants store water in thick stems.
   Some plants have waxy or hairy leaves that keep water from evaporating.
   Some desert plants have deep roots that pull water from far underground.
   Some plants have roots that spread over a big surface to collect water when it rains.

3. Answers could include the following:
   Desert animals are able to live with very little water.
   Some animals get water from seeds and plants they eat.
   Meat-eaters get water from the animals they eat.
   Some animals lick dew off rocks and plants.

Page 123

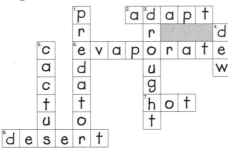

Page 124

Answers will vary.

Page 127

A. Answers will vary.
   1. Recycling is using old things to make something new.
   2. Recycling reduces the amount of trash that is thrown away. Recycling saves resources. Recycling makes less pollution.
   3. Place used papers, glass, cans, and plastic in recycling bins.

B. Answers will vary.

Page 128

1. yard wastes
2. separate
3. pollution
4. trash
5. recycle
6. container
7. products

Page 129

Drawings will vary.

Page 132

1. It is lightweight.
   It is inexpensive to make.
   It doesn't break easily.
   Many kinds of products can be made from plastic.
2. plate—molded
   plastic wrap—rolled
   bottle—blown and molded
3. It is ground up, melted, and made into new products.
4. Answers will vary.

Page 133

Page 134

Answers will vary.

Page 137

A. 1. It is the largest and most complete T-Rex fossil found so far.
   2. Scientists can find out about plants and animals that lived millions of years ago by studying fossils.
   3. eggs, skin, teeth, footprints

B. 3
   5
   2
   4
   1

Page 138

1. T-Rex
2. prehistoric
3. skeleton
4. fossil
5. decay
6. minerals
7. dinosaur
8. paleontologist
9. ancient

Page 139

Answers will vary.